Computational Phenotypes

OXFORD STUDIES IN BIOLINGUISTICS

GENERAL EDITOR
Cedric Boeckx, Catalan Institute for Advanced Studies (ICREA) and Department of Linguistics at the Universitat de Barcelona

ADVISORY EDITORS
Anna Maria Di Sciullo, Université du Québec à Montréal; Simon Fisher, The Wellcome Trust Centre for Human Genetics; Tecumseh Fitch, Universität Wien; Angela D. Friederici, Max Planck Institute for Human Cognitive and Brain Sciences; Andrea Moro, Vita-Salute San Raffaele University; Kazuo Okanoya, Brain Science Institute, Riken; Massimo Piattelli-Palmarini, University of Arizona; David Poeppel, New York University; Maggie Tallerman, Newcastle University

PUBLISHED

The Biolinguistic Enterprise: New Perspectives on the Evolutionary Nature of the Human Language Faculty
edited by Anna Maria Di Sciullo and Cedric Boeckx

The Phonological Architecture: A Biolinguistic Perspective
by Bridget Samuels

Computational Phenotypes: Towards an Evolutionary Developmental Biolinguistics
by Sergio Balari and Guillermo Lorenzo

IN PREPARATION

Language Down the Garden Path: The Cognitive and Biological Basis for Linguistic Structure
edited by Montserrat Sanz, Itziar Laka, and Michael K. Tanenhaus

The series welcomes contributions from researchers in many fields, including linguistic computation, language development, language evolution, cognitive neuroscience, and genetics. It also considers proposals which address the philosophical and conceptual foundations of the field, and is open to work informed by all theoretical persuasions.

Computational Phenotypes

Towards an Evolutionary Developmental Biolinguistics

SERGIO BALARI AND GUILLERMO LORENZO

OXFORD

Great Clarendon Street, Oxford OX2 6DP
United Kingdom

Oxford University Press is a department of the University of Oxford.
It furthers the University's objective of excellence in research, scholarship,
and education by publishing worldwide. Oxford is a registered trade mark of
Oxford University Press in the UK and in certain other countries

© Sergio Balari and Guillermo Loenzo 2013

The moral rights of the author have been asserted

First Edition published in 2013
Reprinted 2013

All rights reserved. No part of this publication may be reproduced, stored in
a retrieval system, or transmitted, in any form or by any means, without the
prior permission in writing of Oxford University Press, or as expressly permitted
by law, by licence or under terms agreed with the appropriate reprographics
rights organization. Enquiries concerning reproduction outside the scope of the
above should be sent to the Rights Department, Oxford University Press, at the
address above

You must not circulate this work in any other form
and you must impose this same condition on any acquirer

British Library Cataloguing in Publication Data
Data available

Library of Congress Cataloging in Publication Data
Data available

ISBN 978-0-19-966547-1

Computational Phenotypes

Towards an Evolutionary Developmental Biolinguistics

SERGIO BALARI AND GUILLERMO LORENZO

OXFORD

UNIVERSITY PRESS

Great Clarendon Street, Oxford OX2 6DP
United Kingdom

Oxford University Press is a department of the University of Oxford.
It furthers the University's objective of excellence in research, scholarship,
and education by publishing worldwide. Oxford is a registered trade mark of
Oxford University Press in the UK and in certain other countries

© Sergio Balari and Guillermo Loenzo 2013

The moral rights of the author have been asserted

First Edition published in 2013
Reprinted 2013

All rights reserved. No part of this publication may be reproduced, stored in
a retrieval system, or transmitted, in any form or by any means, without the
prior permission in writing of Oxford University Press, or as expressly permitted
by law, by licence or under terms agreed with the appropriate reprographics
rights organization. Enquiries concerning reproduction outside the scope of the
above should be sent to the Rights Department, Oxford University Press, at the
address above

You must not circulate this work in any other form
and you must impose this same condition on any acquirer

British Library Cataloguing in Publication Data
Data available

Library of Congress Cataloging in Publication Data
Data available

ISBN 978-0-19-966547-1

To the memory of Richard Owen and Pere Alberch, who we hope would have enjoyed reading this, should they have lived in these days.

Contents

General Preface	viii
List of Figures	ix
List of Tables	x
List of Abbreviations	xi
Acknowledgments	xii

1. Introduction: The Pains of Being Owenians/Chomskyans/Cartesians at Heart — 1
2. My Beloved Monster — 26
3. The Dead End of Communication — 42
4. On True Homologues — 64
5. Computational Homology — 89
6. Introducing Computational Evo Devo — 108
7. Other Minds — 134
8. Conclusions — 161

Appendix: On Complexity Issues	173
References	185
Index of Authors	219
General Index	226

General Preface

This series aims to shed light on the biological foundations of human language. Biolinguistics is an important new interdisciplinary field that sets out to explore the basic properties of human language and to investigate how it matures in the individual, how it is put to use in thought and communication, what brain circuits implement it, what combination of genes supports it, and how it emerged in our species. In addressing these questions the series aims to advance our understanding of the interactions of mind and brain in the production and reception of language, to discover the components of the brain that are unique to language (especially those that also seem unique to humans), and to distinguish them from those that are shared with other cognitive domains.

Advances in theoretical linguistics, genetics, developmental and comparative psychology, the evo-devo program in biology, and cognitive neuroscience have made it possible to formulate novel, testable hypotheses concerning these basic questions. *Oxford Studies in Biolinguistics* will contribute to the emerging synthesis among these fields by encouraging and publishing books that show the value of transdisciplinary dialogue, and which highlight the unique research opportunities such a dialogue offers.

Contributions to the series are likely to come from researchers in many fields, including linguistic computation, language development, language evolution, cognitive neuroscience, and genetics. The series welcomes work that addresses the philosophical and conceptual foundations of the field, and is open to work informed by all theoretical persuasions. We expect authors to present their arguments and findings in a manner that can be understood by scholars in every discipline on which their work has a bearing.

Computational Phenotypes builds on a deep reading of the history of "evo-devo" to argue for a shift away from function in the discussion of the evolution of language. The authors put forward a novel hypothesis about how language emerged in our species that emphasizes the role of brain development and the notion of computation.

Cedric Boeckx

List of Figures

2.1	"...El cirujano no ha querido | finalizar lo comenzado..." ("...The surgeon did not want | to finish what he started...")	30
2.2	"One day the world will be ready for you and wonder how they didn't see."	32
2.3	"Delta Sun Bottleneck Stomp"	34
2.4	"Of Moons, Birds & Monsters"	38
4.1	Guess which xs are the same x (or a different x)	83
4.2	Guess whether the ancestor's part is or is not a homologue	84
4.3	Who's afraid of the archetype?	85
5.1	"Worms vs. Birds"	93
5.2	"Noam Chomsky Spring Break 2002"	95
5.3	"Amsterdam"	99
5.4	"Something Should Be Said"	101
6.1	Guess where your phenotype is	118
6.2	Guess where your phenotype is (from another view)	119
6.3	Guess how your phenotype will slide	120
6.4	Guess how your phenotype will slide (a view from below)	120
6.5	"I've Got You under My Skin"	121
6.6	Guess where your computype is	124
6.7	Climbing Mount Cytogenesis: Guess where you are	128
7.1	"This monkey's gone to heaven."	142
7.2	Screaming trees	144
7.3	"My head's in knots." (I)	147
7.4	"My head's in knots." (II)	148
7.5	"Birdbrain"	152
7.6	The Neanderthal's necklace?	154

List of Tables

4.1	Guess what these behaviors are for	73
6.1	Pick your favorite heterochrony	127
6.2	Pick your favorite mutant	130
A.1	Constraints on rules of grammars	175

List of Figures

2.1 "...El cirujano no ha querido | finalizar lo comenzado..." ("...The surgeon did not want | to finish what he started...")   30
2.2 "One day the world will be ready for you and wonder how they didn't see."   32
2.3 "Delta Sun Bottleneck Stomp"   34
2.4 "Of Moons, Birds & Monsters"   38
4.1 Guess which xs are the same x (or a different x)   83
4.2 Guess whether the ancestor's part is or is not a homologue   84
4.3 Who's afraid of the archetype?   85
5.1 "Worms vs. Birds"   93
5.2 "Noam Chomsky Spring Break 2002"   95
5.3 "Amsterdam"   99
5.4 "Something Should Be Said"   101
6.1 Guess where your phenotype is   118
6.2 Guess where your phenotype is (from another view)   119
6.3 Guess how your phenotype will slide   120
6.4 Guess how your phenotype will slide (a view from below)   120
6.5 "I've Got You under My Skin"   121
6.6 Guess where your computype is   124
6.7 Climbing Mount Cytogenesis: Guess where you are   128
7.1 "This monkey's gone to heaven."   142
7.2 Screaming trees   144
7.3 "My head's in knots." (I)   147
7.4 "My head's in knots." (II)   148
7.5 "Birdbrain"   152
7.6 The Neanderthal's necklace?   154

List of Tables

4.1	Guess what these behaviors are for	73
6.1	Pick your favorite heterochrony	127
6.2	Pick your favorite mutant	130
A.1	Constraints on rules of grammars	175

List of Abbreviations

acc	accusative
CCC	Central Computational Complex
CNS	Central Nervous System
CS	Computational System
ESS	Evolutionary Stable Strategy
Evo Devo	Evolutionary Developmental Biology
EW	Eshkol-Wachman notation
FAP	Fixed Action Pattern
FL	Faculty of Language
FLB	Faculty of Language in a Broad Sense
FLN	Faculty of Language in a Narrow Sense
ka	thousands of years
MAP	Modal Action Pattern
NNSC	Nonnomic Sensitive Creature
nom	nominative
NSC	Nomic Sensitive Creature
OV	Object-Verb language
PNS	Peripheral Nervous System
quot	quotation
SPS	Strong positive selection
t	trace
VO	Verb-Object language

Acknowledgments

This book is the distillation of more than ten years' research we've been carrying out both jointly and independently. It all started well before we came to know each other, but not until this happened did appropriate synergies start to pull in the right direction. For this reason our first acknowledgment goes to Gemma Rigau and Joana Rosselló who, through different means, made our meeting possible and consequently inspired us to start work on the first sketches of what would eventually become this book. In many ways it is also through Gemma and Joana that a fairly loose network of geographically scattered acquaintances came to be a consolidated group of friends and colleagues which includes (in strict alphabetical order): Antonio Benítez-Burraco, Marta Camps, Víctor M. Longa, and Juan Uriagereka, with whom we've been involved in two consecutive research projects on biolinguistics since 2007. Working with them in parallel endeavors while writing this book has been an invaluable source of help and inspiration, and it is difficult to quantify the extent to which their influence has contributed in streamlining, modifying, and refining our ideas. What is certainly undeniable is that thanks to them this work is a great improvement on what it would otherwise have been.

A very special acknowledgment goes to Pere Alberch, whom we never met and only know through his writings, but the reader will soon appreciate that his ideas permeate almost every single line of this book. It is moreover through Alberch that we discovered the work of many eighteenth- and nineteenth-century biologists—most notably Richard Owen, but also Geoffroy Saint-Hilaire, Vicq-D'Azyr, Goethe, von Baer, and many others, and that we learned how cruel history can be when it's written only by the winners: *vae victis*. Our work is, in many senses, a vindication of their thought, unjustly forgotten, ignored, and often perverted by modern biological historiography. In this connection, Laura Nuño de la Rosa deserves special mention for making available to us some material that Alberch left unpublished after his untimely death.

Definitive encouragement to turn what was a fairly heterogeneous and scattered collection of journal articles, conference presentations, and entries in notebooks into a book came from Cedric Boeckx, whose enthusiasm for our work always came as a (highly gratifying) surprise. Were it not for Cedric's commitment to this project, we would certainly not have entered into book-production mode and would still be in the collector's phase.

Some scholars have been so kind as to make available to us papers and articles of theirs we were having difficulties in finding. Very special thanks to Professors Peter Klopfer, Alessandro Minelli, and Georg Striedter, who reacted to the unusual requests of two linguists doing biology by sending both papers and encouragement.

Professor Minelli also spent some of his valuable time reading Chapters 4 and 5, and his detailed comments and suggestions not only reassured us in thinking that our treatment of some delicate biological topics was not complete nonsense, but also helped make it more substantial. We also wish to thank David Hernández for helping us with the artwork of this book and Professor Glenn Tattersall for making available and giving permission to use some images from his private collection. Four anonymous reviewers read the complete manuscript or some parts of it and provided some extremely useful comments and hints on how to improve it. We tried our best if not to follow all their suggestions at least to thoroughly revise and refurbish its chapters in order to turn the original text into something much better than it was before the reviewers' intervention.

Julia Steer, our editor at Oxford, has always been enthusiastic, encouraging, and empowering throughout the whole process leading to acceptance first and production afterwards. For all this, she deserves a big "Thank you, Julia!"—we raise our glasses as a toast. We also want to thank Vicki Hart, Michael Janes, Huw Jones, and Jennifer Lunsford from Oxford University Press for their invaluable help during the preparation of the final typescript.

Finally, thanks to the many bands and musicians who have been good company throughout the process of completing this work. The reader can bet that while most of the pages of this book were being written, some piece of music was playing on the stereo (or similar device) or just echoing in the appropriate brain area of one or the other author. This book, then, like many films, has a soundtrack, and we couldn't resist the temptation of letting some excerpts leak into the text in the form of more or less hidden references. We leave to the reader the choice of discovering them as a complementary exercise of exegesis of what it is like to be Balari and Lorenzo.

Before closing this foreword, we would like to express our gratitude to the institutions from which we received financial support for writing the book: Ministerio de Ciencia e Innovación (Spanish Government) and Fondos Europeos de Desarrollo Regional (FEDER) funds (Balari and Lorenzo; FFI2010-14955/Balari; FFI2010-20634), and Generalitat de Catalunya (Balari; grant 2009SGR1079 to the Universitat Autònoma de Barcelona).

Barcelona, Braga, Oviedo, and Terrassa, Spring 2011–Summer 2012

Picture Acknowledgments

The authors and publishers have made every effort to contact copyright holders of figures reprinted in *Computational Phenotypes*. This has not been possible in every case, however, and we would welcome correspondence from those individuals/companies we have been unable to trace.

La vie est formation de formes, la connaissance est analyse des matières informées.[1]

Georges Canguilhem, 1965

Ogni qualvolta il potere dichiara che la storia è finita e che la natura ha un ordine divino che la regge, sicché felice può essere solo l'uomo che obbedisce e crede, allora il mostro appare a sconfessare ogni normalità, a dire miserabile l'obbedienza e stolta la credenza. Il mostro è un cavaliere che ci trascina in luoghi pericolosi, ma nello stesso tempo ci libera dal dogmatismo e ci incita alla creazione di nuovi mondi.[2]

Toni Negri, 2001

---

[1] "Life is formation of forms, knowledge is analysis of informed matters." (Our translation, SB and GL.)

[2] "Every now and then power declares the end of history and that there is a divine order that rules nature, such that only those who obey and believe can be happy. But then the monster appears, who discredits normality in all its forms, proclaiming obeisance to be despicable and belief to be foolish. The monster is a knight who drags us to dangerous places but who, at the same time, delivers us from dogmatism and encourages us to create new worlds." (Our translation, SB and GL.)

1

Introduction: The Pains of Being Owenians/Chomskyans/Cartesians at Heart

> Der Mensch besitzt die Fähigkeit Sprachen zu bauen, womit jeder Sinn ausdrücken lässt, ohne eine Ahnung davon zu haben, wie und was jedes Wort bedeutet.—Wie man auch spricht, ohne zu wissen, wie die einzelnen Laute hervorgebracht werden.
> Die Umgangsprache ist ein Teil des menschlichen Organismus und nicht weniger kompliziert als dieser.[1]
>
> Ludwig Wittgenstein, 1922

> Mental processes in general and linguistic processes in particular come in two flavors—habits and computations.
>
> David J. Townsend and Thomas G. Bever, 2001

> Après avoir étudié les anomalies dans leurs conditions spéciales, et établi les lois et les rapports généraux auxquels peuvent se ramener tous les faits particuliers, je montrerai comment ces lois et ces rapports ne sont eux-mêmes que des corollaires des lois les plus générales de l'organisation.[2]
>
> Isidore Geoffroy Saint-Hilaire, 1832

This is a book about the Faculty of Language (FL), a technical term that refers to a particular aspect of human cognition and, ultimately, human biology (Chomsky 1980). However, most uses of the word "language" have very little to do with cognition and

---

[1] "Man possesses the capacity of constructing languages, in which every sense can be expressed, without having an idea how and what each word means—just as one speaks without knowing how the single sounds are produced.
  Colloquial language is a part of the human organism and is not less complicated than it." (Translation by C. K. Odgen.)

[2] "Having studied anomalies in their specific conditions and established the laws and general relations underlying all particular facts, I will demonstrate the way in which these laws and relations are but corollaries of even the most general laws of organization." (Our translation, SB and GL.)

biology (Chomsky 1986), so a good terminological choice to refer unambiguously to the target of our investigation is not a minor question for our enterprise. The term FL now has a relatively long tradition, associated with the name and ideas of Noam Chomsky, but even within this tradition it continues to be a rather elusive and problematic concept. Our particular terminological concern is actually twofold.

On the one hand, the term is often seen as controversial because of its inherent association with some notion of "innateness," which Chomsky himself has sometimes used in an ambiguous way, but always somehow connected with the Cartesian tradition (Chomsky 1966). Be that as it may, the term "faculty" had a very definite meaning for Descartes and this is the idea we mean when we use the term:

> This is the same sense as that in which we say that generosity is "innate" in certain families, or that certain diseases such as gout or stones are innate in others: It is not so much that the babies of such families suffer from these diseases in the mother's womb, but simply that they are born with a certain "faculty" or tendency to contract them. (Descartes 1648 [1985]: 303–304— originally written in Latin; translated by D. Murdoch)

For us then, speaking of language as innate is straightforward when viewing it as a faculty in this Cartesian sense (Stich 1975, for an earlier interpretation along these lines). Problems arise when we forget that most of the time, when talking about language, people do not mean "language-as-a-faculty," and that in these more common uses, innate and acquired components are mixed up in the reference of the word "language." It is therefore important to draw as clear a line as possible demarcating what is within the innate component of language—within language-as-a-faculty or FL—and, complementarily, what parts of language are better understood as historical developments socially transmitted within speaking communities. This is a controversial matter and readers may disagree over the point where we locate the divide. However, we consider this a crucial step for our evolutionary proposal to be properly understood. Most of this introduction declares our own vision concerning the extent to which language can be said to correspond to an innate faculty—which, for us, is the same as saying an "evolved" organ.

On the other hand, problems over the notion of FL have to do with questions concerning the extent to which it overlaps with other human or non-human faculties—as witnessed by recent discussions following the publication of Hauser et al. (2002). These authors' main contention is that the notion of FL, as referring to a natural object, needs some further conceptual articulation when used within an evolutionary framework. From this perspective, approaching FL as a particular innate aspect of human cognition/biology is just a first (minor) step in order to move to more substantial questions, namely: Is FL a human-specific capacity? If the answer is "yes:" Is it a domain-specific capacity? If answers are of a "partially so" sort, then: What are the component parts of FL and how do they relate to the rest of human cognition and nature at large? A divide can be established between the

Faculty of Language in a Broad Sense (FLB), comprising the entire set of components of the capacity, and the Faculty of Language in a Narrow Sense (FLN), consisting of just its species and domain-specific components.

A curious consequence of this conceptual divide is that it has created a sociological divide in the field of evolutionary linguistics: There are those who defend the standpoint that nothing in nature seriously resembles FL (Pinker and Jackendoff 2005; Jackendoff and Pinker 2005); and those who argue that very few of the component pieces of language lack a more or less clear correspondence with facets of other minds—or with other facets of the human mind (Hauser et al. 2002; Fitch et al. 2005). The FLB/FLN divide is thus nonsensical for the former, whereas for the latter, FLN is perhaps a residual, but still nonempty, component of FLB and, even if residual, a component with far-reaching consequences, judging from the strong contrast between the cognition/behavior of humans and that of other (narrowly non-linguistic) species.

As for this debate, we belong with what we believe to be a rather marginal extreme position: No single component of FL lacks a corresponding part in the mind/brain organization of other species; or, in the jargon we (Owenians at heart) prefer to use: Every single piece of FL has "homologues" in the brains of other species, irrespective of the level of analysis we choose to adopt. This claim extends to the component that Hauser, Chomsky, and Fitch identify as FLN, i.e., the brain's engine in charge of "recursion," which according to them is the hallmark of language specificity. In a way, we align with those who argue that recursion is not a human-specific feature (e.g., Gentner et al. 2006) or those who contend that it is not a language-specific characteristic (e.g., Corballis 2007, 2011), but for different reasons. In our own view, the system in charge of recursion (not just recursion as an abstract property) subserves many other tasks (not just language) and is most probably a common feature of the brain of vertebrates (not just of humans). Once this claim is justified in this book, the FLB/FLN distinction will be of little use for us.

It is not, we urge to clarify, that we (Chomskyans at heart) consider these questions and concepts nonsensical. They make sense in as much as they have empirical import, meaning that they will ultimately be resolved via experimental observation and testing, and not by conceptual argument, as repeatedly underscored in Hauser et al. (2002). In this book, we will try to fulfill their methodological requirement by supporting our ideas with data coming from the empirical study of human and non-human cognition. We also hope that these ideas could inspire some experimental designs capable of verifying or falsifying them. In this, we will try to approach as much as possible what, we are afraid, is still an ideal state of affairs, far from the current state of the art of evolutionary linguistics, a branch of the emergent field of biolinguistics (Lenneberg 1967; Jenkins 2000; Boeckx and Grohmann 2007; Berwick and Chomsky 2011, among others). We think, however, that this is not something to be unjustifiably worried about. As an emergent discipline, biolinguistics still faces many foundational problems, which explains that the field continues to be dominated by conceptual, rather than empirical, queries. We believe that these queries are not to be avoided and that it is very important to confront and clarify at least some of these conceptual issues before expecting the attainment of real progress in this area of research in the near future. We devote entire parts of the book to these conceptual/foundational tasks. Important as we (Cartesians at heart) consider them to be, we hope to have maintained a reasonable balance between these foundational considerations and other, more empirically oriented, matters.

For the time being, we want to dedicate some space in this introduction to clarifying some basic conceptual questions underlying our approach to language evolution before starting to elaborate our "long argument" in the next chapter. In the first section we situate the readers at the level of linguistic analysis about which the book as a whole tries to offer a plausible evolutionary explanation, and present a detailed justification for the idea that such a level exhausts the biological (or evolved) dimension of language—i.e., FL. Readers will then notice that ours is a concept of FL fairly different from that of many other scholars, in that no specifically linguistic representations and mechanisms exist at this level of analysis. While still thinking that this is through and through a book about FL, it seems to us that this is a strong enough reason to conclude that our argument may benefit from a terminological twist, even at the risk of introducing some further complications into the general picture of the discussions on language evolution. The second section of this chapter introduces the idea that not having so far been able to locate the primary level of analysis for the evolutionary explanation of language has given rise to approaches that misidentify it as a member of ungrounded biological categories. This is the case with the extremely influential "communicative approach" to language evolution, which relies on the category of "animal communication"—a biologically illusory one, as we will argue at length in Chapter 3. The problem with this and similar approaches is that while they trust in observable behaviors as an adequate level of analysis for grounding evolutionary generalizations, the evolved component of language happens to be deeply rooted in the human phenotype. This final section offers the outline of a strategy for accessing such a level of analysis by bridging the gap between this deeper or more abstract level and the one corresponding to what is merely observable; the section also functions as a general overview of the book.

1.1 On Granularity, Commensurability, and Flavors

Two particularly important foundational issues for the field of biolinguistics, raised by Poeppel and Embick (2005), can be jointly referred to as the "granularity/incommensurability" problem. So amalgamated, the problem has to do with the fact that observations coming from theoretical linguistics have a much finer degree of resolution (or "granularity") than those coming from neuroscience and with the fact that the units (or "currencies") routinely used to describe or explain operations in the corresponding fields do not seem to be mutually interchangeable (or "commensurable"). Thus, unification of the fields seems unlikely in the near future, contrary to the desiderata of the biolinguistic community. In particular, the importance of this problem for the progress of evolutionary linguistics has been emphasized by Boeckx (2009) and Hornstein (2009), who rightly observe that FL, as it is commonly understood, looks like a very suspicious evolved phenomenon, given its high complexity and intricacy (judging from the descriptions of theoretical linguists) and the small time span it took to evolve (judging from the paleoanthropological evidence so far gathered; see Balari et al., in press, for a critical review).

Furthermore, as also pointed out by Boeckx (2009) and Hornstein (2009), linguistic analyses are based on categories that are both human and domain specific, thus apparently preventing the application of the comparative method of evolutionary biology. It is true on a trivial level that languages are made up of things without known equivalents in the typical productions of other species—anaphors, pronouns, case and agreement markers, etc., to name a few. So, clarifying language evolution will look like a hopeless enterprise, were it to be solely aimed at explaining the emergence of these curious pieces of linguistic "hardware." This observation apparently points to the somewhat problematic conclusion that language is not such a monolithic phenomenon as we routinely see it; more specifically, it seems to support the view that language is partially made up both of an evolved component and of human inventions. We will refer to this question as the problem of the "dual nature of language."

This book contains some concrete proposals regarding these questions that are not the subject matter of any specific chapter, but transversal themes of the book as a whole, which we will now devote some space to introducing before proceeding with an outline of the different chapters.

1.1.1 Granularity, commensurability, and minimalism

As for the first problem, we have a "modest proposal:" We suggest focusing the evolutionary study of FL on the level of analysis, following Pylyshyn (1984), now referable as "functional architecture." This level corresponds with a set of basic mental operations directly provided by the biological substratum of mind (such as "accessing," "sequencing," "storing," "comparing," and so on), which endow cognition with its essential powers, but at the same time severely restrict the scope of these powers. As an aspect of the biological (i.e., naturally evolved) substratum of the mind/brain, these operations are expected to be subject to a good range of interspecific variation. Within Pylyshyn's cognitive framework, they don't belong to the symbolic code of any particular mental system or module (language, vision, motor control, or whatever) and they are semantically vacuous (i.e., they lack any particular or specific connections with the conceptual/intentional systems of mind). We are in complete agreement with Pylyshyn's picture. However, in this book we take one step beyond and we also contend that these basic or architectural operations are provided by a particular system of the mind/brain: The Computational System. Actually, this is the system we referred to above as possibly being present in the brains of the whole vertebrate phylum and as simultaneously subserving many different tasks in each particular brain; not exactly the same thing we think Chomsky has in mind when he uses the same term—as in Chomsky (1995) and subsequent works.

Momentarily leaving aside the empirical plausibility of this idea, we think its conceptual advantages in the context of the "granularity/incommensurability" problem are clear and various, and so we will try to enumerate them in order:

1. First, we think that this architectural level can be a valid common ground for both theoretical linguists and neuroscientists (as well as geneticists, evolutionary biologists, and so on), in that it lacks the subtleties of linguistic analyses, without omitting a core set of categories which can very reasonably be put in relation to those analyses.
2. It also seems reasonable, whatever the extent and complexity of FL ultimately, in understanding its evolution, to try to begin to explain its minimal cognitive requirements. From these humble beginnings, explanations can progress towards much more ambitious goals.
3. Point 2 admits a reading in which our project can be seen as a variant of Chomsky's minimalism (Chomsky 1995, 2000, 2004, 2005, among other works), in that we can entertain the idea that these minimal properties of the Computational System actually comprise the whole evolved component of FL. This amounts to saying that linguistic computations might be completely fulfilled by means of the resources provided by the functional architecture of a general purpose computational device, with no aid from additional functions and/or symbolic codes dedicated to other language-related or even language-specific tasks—which in any case presuppose those more basic or primitive resources. Were this actually to be the case, we would be facing a good illustration of a maximally efficient cognitive capacity or faculty, considering a hierarchically organized model of the decomposition of functions within the computational brain (Gallistel and King 2009).

With this idea in mind, we can establish something like the following minimalist research agenda:

An evominimalist methodological suggestion

1. Try this: The Computational System (remember, a species- and domain-*unspecific* brain component) provides FL with its whole set of evolved computational procedures;

2. If (1) fails, try this: FL also incorporates evolved computational resources (mechanisms and representations) from the external systems (i.e., sensorimotor and intentional/conceptual modules);
3. If (2) is not enough, try this: FL also incorporates evolved computational resources (mechanisms and representations) of a domain-specific sort.

While our book will be an in-depth exploration of step (1) of this plan, which we argue to be a very promising working hypothesis, we cannot dismiss beforehand the possibility that corrections with the flavor of steps (2) and (3) will be required as more detailed inquiry progresses. For the moment, we just affirm that progression in this field will necessarily grow from the fertile soil of the level of analysis in which, for both methodological and empirical reasons, we have decided to locate our own explanations.

1.1.2 Two flavors of language

We are afraid, however, that the "granularity/incommensurability" problem is not the only foundational problem that biolinguistics needs to clarify before transforming itself into a fruitful discipline. There is another, perhaps even more serious, foundational problem whose eventual solution may in the end help shed some light on the issue of "granularity/incommensurability". This problem, which we will refer to as the "dual nature of language," concerns a pervading confusion, both in linguistics in general and in evolutionary linguistics in particular, between the biological/computational system with which all members of the human species are endowed and the psychological/cultural systems every member of the species is capable of developing—namely, for lack of a better term, "grammars." Thus, as Ludwig Wittgenstein put it in the *Tractatus*, language is, in some sense, a human invention—"grammars," but *also*, and perhaps in a different sense, an integral part of the human organism—FL. This duality, so succinctly exposed by Wittgenstein in his first philosophical work, is perhaps one of the most challenging aspects that any scientific approach to language will ever face, and is, deservedly, one of the central topics of this book. It is a topic that we will try hard to unpack in the chapters that follow, but one to which we would like to devote some space in this one, as it is perhaps the origin of many misunderstandings and of some of the most acrimonious debates in the field of evolutionary linguistics. The issue can be summarized very briefly by stating that it is one thing to investigate the origin of FL (with the specific technical meaning we reserve for this term here) and another, different thing to speculate on the processes, contexts, and contingencies that favored the emergence of grammatical systems. We regard the former as a strictly organic question, whereas the latter concerns the interaction between biology and culture. Our contention here is that the former, the organic aspect of language, is basic—actually, prior—if we want any just-so story about the invention of grammars to make any sense at all, but also if we want to achieve a better understanding of language as a whole. Let us expand this topic further.

1.1.2.1 Language: The evolved barebones and the social flesh Languages—say English, Catalan, Turkish, etc.—are made of apparently endless collections of rules, sometimes difficult to grasp at a conscious level—even for native speakers, and in most cases subject to exceptions and a certain degree of indeterminacy in their application. Furthermore, language-specific rules apply across a wide array of domains, governing the well-formedness of sound patterns, word and phrase structures, and—more controversially (see Chierchia 1998)—logical forms. A wide space for inter-linguistic variation consequently exists—as well as a good justification for the nightmare that learning a new language represents to most adults. Even so, all languages obey a common set of basic principles of design, which—we are going to argue—cannot be conceptualized as just a contingent effect of convergence of the language-specific collection of rules—or grammars—of all existing languages. As a matter of fact, we strongly believe that some of these principles are not even part of grammars and that they are to be better conceptualized as part of a biological substrate underlying the development, knowledge, and use of particular grammatical systems, as well as of other cognitive domains. We have expressed this idea before by saying—following Pylyshyn (1984)—that these principles belong to the "functional architecture" of mind, meaning that they are expected to manifest themselves across different domains of application and to shape the particular representational codes holding in each domain, while keeping themselves autonomous from these domains.

We want now to elaborate on this idea and, especially, to introduce and justify the thesis that the complexity of grammatical codes—and the corresponding mental representations—is rigidly bounded by architectural constraints, although still open to an enormous degree of variation in the particulars of grammatical machineries. In doing so, we also pave the way to another important thesis concerning the proper target of the evolutionary process leading to FL. Our contention in this respect is that the evolution of the suitable computational machinery—which we identify with the above-mentioned component of the functional architecture of mind—has made the historical development of local grammatical conventions possible, which are not—properly speaking—part of the evolved component of FL—an idea we hinted at in earlier work (Balari 2005, 2006) and also recently put forward by Chomsky (2010: 61), who actually relates it to the problem of the enormous range of linguistic variation and the rapidity with which variation has accumulated in a relatively short time. This is not to say that grammatical conventions must be thought of as completely alien to the evolutionary recipe of FL in the species or to the developmental system underlying FL in the individual. In fact, we are going to argue that they probably exist—among other reasons—to scaffold the early development of FL in the minds of children, which—given the Evo Devo orientation of our theses—is like saying that

they could also fulfill a similar role in the evolutionary construction of FL. In spite of this, our idea is that grammatical conventions belong to the traditional inheritance of speaking communities and, even if they are capable of interacting with biological processes—both in developmental and evolutionary time, it is to the same extent as many other environmental inputs—such as temperature on sex determination in many species; see Moore (2001) for details—which exert similar influences without being a part of the evolved phenotype in any relevant sense.

1.1.2.2 Three layers of computational complexity Let us start by pinpointing three different layers in the organization of linguistic expressions, named after three linguists who are known for their insights regarding one or another of these strata. This is a paradoxical decision, because our interest in identifying these layers is to show that the tasks underlying them—corresponding to different levels of computational complexity—are assisted by non-linguistic resources. In spite of certain recent controversy—see Everett (2005, 2009) and Nevins et al. (2009a, 2009b), languages are identical at this level of analysis, in that no layer seems to be absent from any of them. If—as we are going to argue—this property is the reflex of more basic—or architectural—properties of the human mind, inextricably connected to the functioning of language, then uniformity at this level is an expected and well-motivated fact, not requiring further and more sophisticated explanation. Following the increasing degree of complexity of the corresponding strata, we introduce in order what we call the *Saussurean*, *Tesnièrean*, and *Chomskyan* layers of linguistic organization. Linguists put them into historical focus, but our enterprise is aimed at justifying the idea that they are grounded on a level of cognitive analysis connected to, but independent from what specifically governs the construction of linguistic expressions.

(i) The Saussurean layer

Ferdinand de Saussure (1916), to begin with, observed that linguistic units are linearly arranged in discourse and proclaimed that—even if obvious and simple—this fact deserved to be stated as one of the most basic or defining features of linguistic systems:

Dans le discours, les mots contractent entre eux, en vertu de leur enchaînement, des rapports fondés sur le caractère linéaire de la langue, qui exclut la possibilité de prononcer deux éléments à la fois. Ceux-ci se rangent les uns à la suite des autres sur la chaîne de la parole. (Saussure 1916 [1955]: 170)[3]

---

[3] "Words as used in discourse, strung together one after another, enter into relations based on the linear character of languages. Linearity precludes the possibility of uttering two words simultaneously. They must be arranged consecutively in spoken language." (Translation by Roy Harris.)

We identify for this reason as "Saussurean layer" units arranged in linear order, as represented in (1):

(1) he > likes > red > apples

Another insight from Saussure is that he also observed that this basic principle of discourse organization was not an arbitrary one: For him, it was clearly motivated, imposed upon languages by the system of exteriorization that they normally choose—i.e., the vocal-auditory channel; see Saussure (1916: Part I, Ch. 1, § 3). This is an observation with far-reaching consequences, but curiously enough, somehow contradictory with the significance that Saussure attributed to linear order as being a primitive or defining principle of languages. Many languages exist—the so-called "polysynthetic languages;" see Baker (1996)—in which grammatical rules concerning the linear order of units—above the word level—do not exist at all—or very marginally; furthermore, in languages in which basic formats of linear order do exist—e.g., VO versus OV languages, such formats do not intrinsically regulate the linear order of units—words or phrases, but rather the linearization of hierarchical relations in discourse—to which we turn in a while. Observations like these invite one to conclude that linear order is grammatically under-regulated—or not regulated at all, most probably due to the fact that it is a feature of the motor systems with which FL interfaces, rather than a property of FL's computational core. A conclusion, by the way, that gets stronger support from the fact that sign languages do not necessarily follow a strict linear pattern and may thereby violate the Saussurean constraint on the non-simultaneity of units.

(ii) The Tesnièrean layer

Lucien Tesnière agreed with Saussure that linear order was the raw material of discourse, the primary datum of language as manifested in speech. However, in the posthumously published book for which he is mostly remembered (Tesnière 1976), he qualified Saussure's insight by saying that, the immediateness of linearity to observation notwithstanding, it was a secondary or parasitic principle relative to what he considered to be the first or nuclear principle of linguistic organization: Linguistic units are primarily arranged as hierarchically organized groups or sets— each set constructed around a particular head or nuclear element and potentially a constituent of a bigger or higher-level set. This principle of structure building is illustrated in (2)—which translates not literally, but rather faithfully, Tesnière's own arboreal representations:

(2) [$_{likes}$ he likes [$_{apples}$ red apples]]

Our main concern here regarding this "Tesnièrean layer" of linguistic organization has to do with the cognitive resources that make it possible to begin with. Note that it is the combined presence of the Saussurean and Tesnièrean layers that makes human languages at least context-free; see Pullum (1986). It is only the presence of the additional layer we are about to discuss that locates them beyond context-freeness. Thus, and leaving aside here many technical questions concerning the formal properties of "Tesnièrean sets"—Is there always a head that projects? Are sets always binary?, etc.—that are irrelevant to our discussion here, let us concentrate instead on the barebones of structures like (2)—as depicted in (3), which is closer to what we have in mind when we talk about this particular layer of organization:

(3)   [x x [x x]]

An important property of this principle of organization is that it goes together with a capacity for relating units at long distances, as it becomes clear when a set is included in another creating a discontinuity in the container set, as in (4):

(4)   [the boy declared [that he likes red apples] in a loud and clear voice]

Just for the sake of the point we want to make here, we can simplify a little and represent the schema of this structure as in (5), the critical parts of which are the sub-indices that we have made use of as indicators at the point at which the major structure is interrupted and at the point at which it is resumed:

(5)   [x x_i [x x x x] x_i x]

This representation is a useful illustration of the main fact that we want to emphasize, namely that what we are referring to here as the "Tesnièrean layer" of organization is the linguistic reflex of a particularly overdeveloped system of operative or working memory with which the human mind is endowed. Such an evolutionary outcome allows the processing of a given task to be interrupted, to keep it frozen in short-term memory—and then resumed and concluded, and for similar subtasks to be processed in the meantime. For us, this amounts to saying that "nested embedding"—the property of Tesnièrean structures illustrated in examples (3) and (5) but, above all, a distinctive feature of any system abstractly characterizable in computational terms as being context-free—is not a piece of the grammatical equipment of FL, but a crucial feature of an underlying computational substrate that—as we argue at length at different points of this book—is also put to work in other cognitive domains—i.e., it is not language specific—and variants of which are also present in other non-human organisms—i.e., it is not human specific.

We commented in passing before that—if regulated at all—what grammatical rules concerning the "Saussurean layer" actually dictate is not the linear order of specific items, but hierarchical relations—as when we classify English as a "head-initial" language (6a) and contrast it with Turkish, a "head-final" language (6b):

(6) a [_write_ write [this book]]
    b [_yaz_ [bu kitabı] yaz]

This is a first illustration of what we view as a true "grammatical convention"—i.e., a historically fixed and socially transmitted habit connected to FL. Note that, in this particular case, the grammatical convention complies with a constraint for which the natural substrate of FL leaves sufficient room for variation, since a hierarchical relation is equally compatible with the two linear arrangements—see Gazdar et al. (1985) for an earlier implementation of this idea, and Boeckx (2010) and Richards (2008) for some recent considerations on the issue.[4] So, what we have here is a situation in which grammar fills in an indeterminacy of nature with a social convention. As we immediately explain, not all grammatical conventions respond to this pattern, but this is a good example to help understand the distinction between the evolved and the historical components of languages—or FL versus grammar—that we want to elaborate on further in this section.

(iii) The Chomskyan layer

The complexity of linguistic expressions, however, does not stop with the "Tesnièrean layer." We do not need to construct a new example to verify it. Let us go back to example (4) again, but noting now that other parts of this expression are also connected in addition to those at the edges of the discontinuous constituent. Again for the sake of simplicity, we are going to concentrate here only on one of these connections, represented in (7) by co-indexation:

(7) [the boy$_j$ declared [that he$_j$ likes red apples] in a loud and clear voice]

The point that we specifically want to emphasize concerning this new kind of long-distance relation is that, in cases like (7), the connection is established by crossing another relation—often more than one—of the long-distance type—in the case of (7), the one signaling the interruption of the processing of the main set. Turning to our reductive technique, we can assign to expressions like (7) the structural skeleton depicted in (8)—where, for the sake of clarity, the new long-distance relation is represented by superscripts:

(8) [x^j x_i [x x^j x x] x_i x]

It is this type of structure with crossing long-distance relations which illustrates what we mean by the "Chomskyan layer" of linguistic organization, as it serves to assign an upper level of computational complexity and locate FL within a particular area within

---

[4] To be fair, this insight is perhaps originally due to Haskell Curry and his notions of "phenogrammatical" and "tectogrammatical" structures, roughly corresponding to our Saussurean and Tesnièrean layers—see Curry (1961).

the Chomsky Hierarchy of languages—Chomsky (1959b, 1963). Note that processing a structure like (8) implies more powerful cognitive resources than processing a structure like (5), since the former requires keeping frozen in working memory space sets of sets of strings until the different crossing relations are resolved. We leave the technical details until Chapter 5; for now, suffice it to say that what makes this "Chomskyan layer" of linguistic organization possible is not a grammatical device, but a further evolutionary overdevelopment of the same cognitive machinery—working memory—that we previously made responsible for nested embedding. This amounts to saying that the "Chomskyan layer" is, again, a linguistic reflex of an evolved basic computational substrate of the human mind—connected to FL, but autonomous of it. This is not to deny the existence of grammatical conventions that work in close connection to this layer. They do indeed exist, but what we are trying to emphasize is that they belong to a different, non-organic level of linguistic analysis. Some additional examples may be useful to stress this point.

Let us consider a well-known cross-linguistic contrast between languages that possess—or do not possess—certain units capable of entering into "Chomskyan relations" of the sort described above. Japanese, for example, possesses an anaphoric unit—*zibun*—that may co-refer to a nominal element located outside its own "Tesnièrean set;" similar anaphoric—i.e., obligatorily co-referential units—in other languages like English—*her-/him-/itself*—are on the contrary disallowed to establish this type of long-distance crossing relations, as the following contrast illustrates:

(9) a [John$_i$-wa [Peter-ga zibun$_i$-o nikunde-i-ru to] omotte-i-ru]
John-topic Peter-nom self-acc hate-asp-pres quot believe-asp-pres
'John believes that Peter hates him.'
b [John$_{*i}$ believes [that Peter$_j$ hates himself$_{*i/j}$]]

As is shown by the indices in (9b), the anaphoric unit in English must be bound within its own "Tesnièrean set," which corresponds to a different grammatical convention from that witnessed in the Japanese counterpart (9a). For us, contrasts like this between Japanese *zibun* and English *himself* show, on the one hand, how grammatical conventions are universally shaped by the constraints imposed by the organic computational basis of FL; but also, on the other hand, that they consist of local and extremely variable systems of regulated units, two marks of their cultural origins and social transmissibility. It goes without saying that the rules governing the use of grammatical items are much more nuanced than a single example can show. In the case of *zibun*, for example, there also seem to be at work a plethora of extra conditions—e.g., it must be topic oriented, the topic must refer to persons, there must be awareness on the topic's side of the attribution of the propositional content, and so on; see Kuno (1973) for details. But we think that the highly convoluted character of grammatical conventions is a further clue to their social roots. A consequence—and we think an important one—of adopting this point of view is that what is normally taken as the grammatical content of such units—customarily formalized as matrixes of features selected from an innate inventory: "[±anaphoric]," "[±long distance]," and so on—is better off simply being understood as the "conditions of use" of the relevant items—in a strictly Wittgensteinian sense,[5] which given a robust organic computational background and a suitable sensitizing period are as easy to grasp as any other social convention. We explore below further consequences of this perspective. For the time being, let us introduce some more examples in order to make even clearer the extent to which we think that languages, even if naturally shaped, are made up of social conventions.

Another sentence from Japanese can be useful to illustrate our next point:

(10) [sono hon-o$_i$ Hanako-ga [Taroo-ga t$_i$ katta to] omotte-i-ru]
      that book-acc Hanako-nom Taroo-nom bought quot believe-asp-pres
      'That book, Hanako believes that Taroo bought.'

This example illustrates another way—different from that of (4)—in which "Tesnièrean sets" can surface discontinuously in sentences. In this case, a constituent—*sono hon-o*—is displaced from its original set—that headed by *katta*, creating a long-distance relation that crosses it. In cases like this, we represent this relation as holding between the displaced constituent and a trace left behind within its original set—"t." The relation is not different from that in (9a) between the long-distance anaphor and its antecedent, but now holding more abstractly between a set and one of its constituents located at long distance.

However, the aspect of (10) that we are most interested in is that the displaced element carries with it a hint of its belonging to a set different from that in which it surfaces: Namely, the accusative marker *-o*. It is tempting to think that elements like these—case, but also agreement markers and similar devices—smooth the task of establishing long-distance relations in complex Tesnièrean/Chomskyan configurations—see Uriagereka (2009) for a congenial approach. This is not to say, however, that this is *the* function of these grammatical devices—as, after all, they also show up when the corresponding sets are continuous. Our contention, once again, is that these pieces of grammatical machinery are historical inventions, which most probably do not obey a single and well-defined cause for existing, but reflect complex and nebulous pressures on speakers that have ultimately to do with

---

[5] The idea is also reminiscent of Émile Benveniste's (1970) distinction between the "use of forms" and the "use of language"—with the former referring to the speaker's knowledge about the conditions of use of particular items, and the latter to her knowledge about the conditions of use of language at large within different social practices. Translating the distinction into Wittgensteinian terms (Wittgenstein 1953), the former connect to different "language-internal games"—phoresis, agreement, case marking, etc.—in a sense not very different from that in which the latter are said to connect to different "language-external games"—such as telling stories, giving instructions, praying, etc.

the computational sophistication of the evolved component of language—FL.[6] The extremely wide range of diversification of case, agreement, or adpositional systems—and combinations thereof; see Blake (2001) and Corbett (2006)—is very reasonably to be taken as a clue to their origins being products of pure historical change (Chomsky 2010: 61), which is not to deny that they nevertheless conform to the natural mold imposed by the cognitive background governing the building of linguistic expressions.

Case and agreement systems have lately been the focus of considerable interest because they enter into the province of so-called "non-interpretable" linguistic devices—Chomsky (1995) and subsequent works, since they do not constructively take part in the thoughts expressed by linguistic expressions. This purports to saying that they are unmotivated from the point of view of the Systems of Thought; if anything, they only seem to introduce further complications to the motor realization of these thoughts. Chomsky's first move concerning this machinery within the Minimalist Program was to deem it as the grammar-specific evolved component of FL—or Universal Grammar, a purportedly residual part of language fixed within the human genotype; see Chomsky (2000, 2004).[7] However, by categorizing it now as the outcome of historical processes, we think that "non-interpretable" grammatical conventions are rightly connected to a different dimension of the design of languages: Namely, the environment—or "Second Factor" in Chomsky's (2005) minimalist parlance.

Our own interpretation of the phenomenon is that grammatical conventions of this "non-interpretable" sort are just part of the external stimuli used by or directed to the speaker/hearer (Lorenzo and Longa 2011). It is obvious that she has to get familiar with them and to grasp the corresponding conditions of use, but we contend that this is so in a way no different from those of other social practices and abilities. After all, we know about "these things" because they are "fully parsable at the level of speech" (Uriagereka 2009: 178), so the most reasonable move—especially from a minimalist oriented prism—is to think about them as given in the external input as their primary substrate. Furthermore, according to our interpretation they work as a sort of "external scaffolding"—in the sense of Clark (1997), the presence of which in the environment excites and strengthens the development and functioning of the

---

[6] Or, perhaps more precisely, they are accidents, albeit not totally unexpected ones, due to the interplay of fairly general grammaticalization trends with the computational constraints imposed by FL; see Heine and Kuteva (2007) for a detailed presentation of these grammaticalization trends and some examples of the emergence of case and agreement systems and other grammatical devices in different languages, and Kiparsky (2008, 2011) for arguments in favor of the idea that grammaticalization is the natural consequence of the action of internal constraints and not, as it is often assumed, of some functional principle acting for the optimization of communicative acts.

[7] Thus corresponding to alternative (3) within our *Evominimalist methodological suggestion* above—i.e, the alternative to be maximally avoided.

complex cognitive activities underlying the exercise of language.[8] This would be a position not so far from that defended in Chomsky (2000)—where the suggestion is made that imperfections help the Computational System in the processing of complex configurations, were it not for the fact that, for us—and it seems that more recently also for Chomsky (2010), it does not follow from this that they belong to the evolved or innate component of language. Rather that they belong to E(xternal)-language—in Chomsky's (1986) terms.

1.1.2.3 A terminological suggestion (and some related issues) Summing up our position so far, we contend that evolutionary linguistics has traditionally failed in establishing a clear-cut distinction between the evolved and the historical components of language. In addition, we believe that grammatical conventions are generally products of history—i.e., locally created and socially transmitted habits associated with the practice of speech, so we must tell them apart from the biological—i.e., naturally evolved—substrate that makes possible the acquisition and use of these conventions. We are perfectly aware that in sustaining this position we depart from most versions of the concept of FL, in which both basic architectural properties of languages and language-specific categories and features are equally deemed innate and thus part of "language-as-a-faculty," an evolved component of the human mind. So in order to avoid confusions it is perhaps a good option to try a new terminology to help readers not to think of the language-specific connotations of more traditional FL concepts—while still understanding that this "grammar-free" concept is all that the concept of FL amounts to according to our point of view. So we suggest using the term "Central Computational Complex" (CCC) as an alternative to FL, such terminological choice being based on the proposal that it simply consists of a computational system—as sketched in 1.1.1, plus a collection of interfaces to the peripheral or external systems that it subserves. An additional advantage of the term is that it also frees the concept of FL from its human-specific connotations—in agreement with the position that we advanced at the beginning of this introduction.

The relevance of this decision will be evident as we unfold our argument. For the time being, let's pause to recall what belongs—and what does not—to this biological level of linguistic analysis, which we render as the proper or primary target of evolutionary linguistics. On the one hand, it definitely belongs to it the capacity of (a) internally sequencing items; (b) keeping sets of items in active memory in order to relate the current sequence with them at a long distance—as in (5); and (c) keeping

---

[8] Our contention is that all this non-interpretable grammatical stuff is a particular illustration of the idea that "evolution has favored on-board capacities which are especially geared to parasitizing the local environment so as to reduce memory load, and even to transform the nature of the computational problems themselves" and that language, therefore, is one of those cases "where evolution has found it advantageous to exploit the possibility of the environment being in the cognitive loop" (Clark and Chalmers 1998: 11). See also Wilson's (1994) "wide computationalism" concept as particularly congenial to our idea.

sets of sets of items in active memory in order to define crossing relations among them—as in (8). For (a) to be carried out a "repetitive sequencing engine" is needed, while operations of the likes of (b) and (c) require a "working memory space" with powers higher than that of a pushdown automaton. In Chapter 5 we present our hypothesis concerning the natural instantiation of such a computational system—as well as clarifications on these computational technicalities. Also belonging to the biological substrate of language are the interfaces connecting this natural system of computation with an open array of peripheral or external systems, so sequences are apt to codify complex thoughts as well as motor and perceptual instructions—both strung together or independently, as well as to subserving many other typically human tasks not customarily seen as properly linguistic—say, painting, manufacturing tools, making knots, and so on, which reasonably depend upon the same system of computation and connections, as we will argue. On the other hand, such properties as being a "head-initial" (English) versus a "head-final" language (Turkish) or having "long-distance anaphors" (Japanese) versus not having them (English) definitely do not belong to the biological substrate of language. This does not purport to saying that humans are not biologically endowed to capture such language-specific features; it simply boils down to the idea that it is just the endowment to capture and use them that belongs to the biology of language, with the exclusion of the features themselves—which rather belong, with the rest of grammar, to a set of linguistic traditions, unfolded in historical time and by sociocultural means.

So note, in concluding this terminological remark, that our FL concept falls short of what many readers would expect as inextricably connected to it—language-specific stuff, such as categories, features, types and conditions on rules, and so on; but, at the same time, it also exceeds what many would accept as a language-dedicated capacity—as we will contend that it also subserves such disparate activities as drawing or tying knots. Taking both sides of the question into consideration, it seems a salutary strategy to use the CCC term when required to prevent misunderstandings from now on.

Having thus established the limit between biology and culture in the realm of language, it is now time to clarify that these are not however enclosed domains. Grammatical conventions could simply not exist were it not for the prior existence of a biological system with the properties of CCC, which is the natural foundation of key aspects of human cultures—technology, art, etc., in addition to linguistic traditions, of course. But, seeing things from the opposite perspective, substantial parts of grammatical systems can be reasonably thought of as fulfilling roles that have to do with the development and functioning of CCC, as hinted at in the previous section. A tentative suggestion regarding this question is that this is probably a primary role of those grammatical devices that have nothing at all to do with the content of thoughts that humans express linguistically—for example, agreement and case systems. They are probably there for their impact on the early development of memory requirements and the online performance of the complex computational tasks implied in the processing of linguistic expressions. For us, the former (developmental) role is maybe the one that most readily explains the historical development of grammatical conventions as traditions inextricably associated with the knowledge and use of languages.

There exist well-documented cases in which certain phenotypic features of organisms are there mostly—or simply—because of the role they had in development—Minelli (2003), for example, defends the argument that the cuticle evolved in many types of organisms as a means of stabilizing shape during development. We adhere to this "primacy of the developmental role" thesis (Minelli 2003: Ch. 2) and, in a similar vein, we argue that apparently unmotivated grammatical conventions are there mostly—maybe simply—because of the role they have in the early development of the evolved cognitive underpinnings of language. A difference exists, of course, in that development is scaffolded from the inside in the former case, and from the outside in the latter. In both cases, moreover, features can afterwards adopt new roles—such as serving as a protective device, in the case of the cuticle, or helping the online processing of expressions, in the case of grammatical conventions.[9]

We could make the comparison more extreme by saying that grammatical conventions are part of the human cognitive phenotype in an extended sense—that of Dawkins (1976, 1982). However, we think that this move would purport to granting grammatical conventions an adaptive value that we do not believe they have—see Balari (2006). It is extremely doubtful that particular case or agreement systems—or improvements thereof—have historically thrived because of their own inherent qualities, instead of simply spreading for strictly exogenous reasons. Furthermore, it is equally doubtful that these systems can evolve to reach some local optima within an adaptive landscape—as it is argued in the case of other social practices; see Avital and Jablonka (2000). Indeed, there exist very good reasons to think that the opposite is true and that nothing (or very little) in linguistic change can be attributed to such principles as optimization and adaptation to improve communicative efficiency. Thus, for example, William Labov in his detailed study of the internal factors participating in linguistic change concludes:

Given phonological and morphological variation, the functional hypothesis predicts a tendency for speakers to choose one variant or the other in a fashion that will preserve information. Most of the results cited show the opposite: in the stream of speech, one variant or the other is chosen without regard to the maximization of information. On the contrary, the major

---

[9] Townsend and Bever (2001) offer a particularly interesting model to implement this idea on the side of comprehension, with "grammatical stuff" serving to inductively inspire the generation of early drafts of structures, to be fully parsed—and eventually corrected—at a later stage of processing by proper "computational procedures." See also Townsend and Bever (2001: Ch. 9) for some reflections on the implications of the idea on acquisition.

effects that determine such choices are mechanical: phonetic conditioning and simple repetition of the preceding structure. (Labov 1994: 568)

Similarly, in a later volume (Labov 2001), he dismisses the traditionally accepted wisdom that biological and linguistic evolution are comparable:

But it is not merely the absence of evidence for evolutionary adaptation that runs counter to Darwin's argument for natural selection. The almost universal view of linguists is the reverse: that the major agent of linguistic change—sound change—is actually maladaptive, in that it leads to the loss of the information that the original forms were designed to carry. (Labov 2001: 10)

In fact, Labov (2001), in order to stress the radically maladaptive nature of change goes as far as stating that "[t]he view of language change as pathological is not mere rhetoric" (Labov 2001: 12). For us, the so far only reasonable assumption is that grammatical conventions are very rigidly confined within the limits imposed by the cognitive resources that capture and process them; a very broad space for variation is open within these limits, but of no real evolutionary significance. This is not the same thing as denying that the historical appearance of grammatical conventions could have had an evolutionary impact on the strengthening and stabilization within populations of the memory working capacity underlying FL—probably in parallel, as we will argue in this book, with mutations occurring at the genetic level. We think that this is a very likely conjecture, which will gain further credibility if the developmental role of grammars in relation to the said capacity gets independently confirmed.[10]

We have gone into some detail in depicting our own stance concerning these issues in order to help the reader pinpoint it in the geography of contemporary approaches to the evolution of language. Thus, we understand that our approach is compatible with the work of Simon Kirby and his collaborators in the sense that they are mostly concerned with implementing computational models for the origin and the evolution of grammars (Kirby 2002; Kirby et al. 2007; Smith and Kirby 2008), or also with the mathematical modeling of language change carried out by Partha Niyogi and Robert Berwick (Niyogi and Berwick 1995, 1997a, 1997b, 1997c, 2009). In particular, we share with Kirby and his collaborators the idea that the traditional approach to language evolution—as represented by the likes of Steven Pinker, Paul Bloom, or Ray Jackendoff; Pinker and Bloom 1990, etc.—centered on natural selection as the main explanatory mechanisms is problematic (see Chapter 6) and that other approaches,

---

[10] Note, in passing, that this "developmental scaffolding" perspective is not incompatible with Labov's (2001) "pathological view" of linguistic change, since Labov's emphasis on the maladaptivity of changes concerns their inconsistency with the conception of language as an instrument of social communication and, hence, with the purported optimization of the transmission of information that would explain the occurrence, transmission, and stabilization of changes. Indeed, Labov's investigations of these issues rather suggest that something quite the opposite from optimization for communication is actually going on; see Labov (1994: Chs. 19–21, 2001: Chs. 13–16, 2010).

far from traditional adaptationism, may be much more successful. In particular, one of their most interesting conclusions, and one which will be of some importance in the chapters to follow, is that "the traditional transmission of language means that the most likely strongly constraining language faculty will not be a language faculty in the strict sense at all, but a more general cognitive faculty applied to the acquisition of language" (Smith and Kirby 2008: 3600). In fact, the authors add: "Any strong constraints from the language faculty are likely to be domain general, not domain specific, and owe their strength to selection for alinguistic, non-cultural functions" (Smith and Kirby 2008: 3601).[11] Thus, as we tried to show above in our brief analysis of some of the structural/grammatical properties of languages, we believe that it is certainly possible to tell these "strong constraints from the language faculty"— CCC—apart from other design features—grammatical devices—which eventually emerged through the interaction of different factors during the process of cultural transmission that resulted in the development of particular languages as we know them today. Interestingly, Smith and Kirby (2008: 3601) conclude that "the uniqueness of language arises from the co-occurrence of a number of relatively unusual cognitive capacities that constitute preconditions for the cultural evolution of these linguistic features," but remain agnostic as to what these preconditions might be, while at the same time consider the task of explaining the evolution of such preconditions a crucial task for evolutionary biolinguistics:

> We have at present no answer to these questions, but we suspect that exploring the evolution of the preconditions for the cultural evolution of language is likely to be more profitable, and more amenable to the comparative approach, than seeking to establish the evolutionary history of a strongly constraining, domain-specific and species-unique faculty of language.

This is precisely our main goal in this book: Namely, to present an evolutionary scenario of the processes that lead to the fixation of these preconditions—CCC—for the cultural evolution of linguistic traditions—grammars.

We are perfectly aware that many readers may disagree with the point at which we locate the nature/nurture divide in the case of language, especially those willing to accept that FL is a much more self-contained and specific module of the human mind, associated with a rich collection of innate categories. And it is probable that this entire book will not be enough to convince them to change their minds. However, we entreat these readers not to abandon us too promptly and to read the book as if it were focused on a particular layer—maybe a very basic and thin one—of the natural composition of language. This is an alternative way of making sense of this book.

---

[11] Interestingly, Labov (1994: 598) maintains a very similar view: "We should not be embarrassed if we find that systemic readjustments in the syntax and the morphology of language are governed by the same cognitive faculty that governs the social behavior of mallard ducks."

1.2 A Further Problem (and an Overview of the Book)

It should be easier to understand now why we find the traditional approaches to the evolution of language extremely problematic. The main underlying problem is, in our opinion, their uncritical commitment to a "communicative approach" that sees language as a particular instance of a larger class of systems of communication. Within a biolinguistic perspective, this position translates into the contention that language is a particular "animal communication system," a natural class comprising an extremely heterogeneous set of behaviors based on the exchange of external signals between organisms (Hockett 1960; Hauser 1997). This approach is grounded on a very superficial analogy and fails to locate the proper level of analysis of language evolution deeply embedded in the human cognitive phenotype, as we argue at length to be the case.

Putting aside some previous criticisms concerning this tenet (Chomsky 1968, 1980), our main objection to the communicative approach is that it drives the evolutionary study of language to a dead end, because the concept of "animal communication system" does not refer to a true natural class and, thus, no contentful evolutionary generalization can be based on it. A detailed elaboration of this argument is the subject matter of Chapter 3 of this book ("The Dead End of Communication"), in which we also trace the history and present the state of the art of the approach, as well as an interpretation of its extraordinary impact on contemporary linguistics and, more specifically, on the evolutionary study of language—see Bickerton (2009), for a recent and representative illustration.

There is, though, a non-negligible aspect of the communicative approach worthy of serious consideration, namely that the evolutionary study of language, as is also the case with any other aspect of natural evolution, is to be framed within a comparative background. But taking into account this basic premise of evolutionary thinking, we discover another crucial failure of the communicative approach, in this case connected to a very problematic stance pervading biological theory at large: That of establishing homological relations using behavioral and, ultimately, functional criteria—see Hall (1994) and Love (2007), for some discussion. In Chapter 4 of this book ("On True Homologues") we carefully elaborate the argument that homologies can only be established on structural (Owenian) grounds and introduce the idea that structural/Owenian homologies (or, plainly, homologies) can be established at several hierarchical levels of structural analysis (Roth 1991; Hall 1994; Wake 2003), to which behavioral analyses clearly do not belong.

In Chapter 5 ("Computational Homology"), we introduce the crucial tenet that one of these structural levels is the "computational level" and present the concepts of "computational homology" and "computype," the latter complementary to other concepts familiar to biologists, namely, those of "zootype" and "phylotype" (Slack et al. 1993; Minelli and Schram 1994; Minelli 1996). With all this well-motivated apparatus in hand, the thesis is presented that the human Central Computational Complex (CCC_{HUMAN} or FL) can be homologized with an extremely large natural class of computational systems, most probably extending to the whole vertebrate phylum, no matter what behaviors these systems underlie nor the benefits they provide in each particular case. Then we offer a detailed presentation of the architectural organization of the organ that we identify as the Computational System of CCC, as well as a proposal concerning its anatomical basis, with a basic distinction between a sequencer/basal component and a working memory/cortical component, partially relying on previous work by Lieberman (2006) and Aboitiz and coworkers (1997, 2006), among others. Putting all these pieces together, a radically different view from that of the communicative approach arises concerning the evolution of FL, to which we refer as the "computational approach."

Once established that the evolution of CCC_{HUMAN} is a particular episode of the historical diversification of natural systems of computation, a chapter follows (Chapter 6: "Introducing Computational Evo Devo") in which we introduce current ideas in the field of Evolutionary Developmental Biology—or Evo Devo (Hall 1999; Carroll 2005; Minelli 2007) and justify why we think that it is the right framework to understand the evolution of cognition and, specifically, of CCCs. We adhere to a particular view on Evo Devo which, combining proposals from Complex Dynamic Systems Theory (Thelen and Smith 1994; Kelso 1995), Developmental Systems Theory (Oyama 2000a; Oyama et al. 2001), and, crucially, the concept of Morphological Evolution (Alberch 1980, 1989, 1991), explains the development/evolution of phenotypes as trajectories leading to particular points within a discontinuous space of possibilities (a "morphospace") constrained by a number of parametric factors—genes being one among equals. Also in this chapter we apply this model to the evolution and diversification of computational systems. We contend that the Chomsky Hierarchy, measuring the relative complexity of different types of "languages"—in the technical sense of formal language theory; Chomsky (1956a, 1959b, 1963), can be directly translated into a "computational morphospace," the parametric factors of which describe the system underlying the development of the organ described in Chapter 5. We argue that, at certain critical points, minor perturbations affecting these factors can bring about major changes in the attained "computational phenotype," and we speculate on the specific perturbations that could be at the base of the "emergence"—in the sense of Reid (2007)—of the Type 1 that defines CCC_{HUMAN}. In our explanation, we adhere to a heterochronic model (Gould 1977) and argue that the process can be described as the result of a coincident set of phenomena of "late termination" in the development of the working memory/cortical component of the organ of computation. In a sense familiar to developmental biologists, we contend that the Computational System of CCC_{HUMAN} is a monstrous outcome (or "teratology") of an otherwise rather common feature of vertebrates. We also devote some space to explain how it became possible for this system to evolutionarily connect with the external systems that also define CCC_{HUMAN}.

Finally, in Chapter 7 ("Other Minds") we offer what we consider a very important consequence of the computational approach, actually, a prediction that we hope future research will fully prove to be empirically correct: Minds other than human are endowed with a system of computation homologous to that of FL in terms of computational complexity. This contention is somewhat controversial for two different reasons. First, it is not a self-evident question that a certain degree of computational complexity represents a criterion for establishing homologies. Homology between some non-human CCC and CCC_{HUMAN} is likely, but only reasons to be spelled out in Chapters 5 and 6 will make clear that it can be established in terms of the activity (and complexity) of the corresponding systems. Second, no such system has ever been devised by the communicative approach, with the consequence that FL continues to be seen within this paradigm as an exceptional character with no clear connections with the rest of the organic world. We contend that this is a failure of the communicative approach that directly derives from the limitations and the artifactual character of the concept of "animal communication," and we offer some preliminary observations pointing to the conclusion that the system of computation underlying certain animal capacities, unrelated to the externalization/exchange of signals, seems to share with CCC_{HUMAN} the complexity of a system of computation of Type 1. A detailed analysis of knot tying in the construction activities of some species of birds (Collias and Collias 1962; Hansell 2000) will serve as our main source of evidence. We believe that this is an aspect in which a computational model like ours demonstrates its predictive power and its capacity to be experimentally tested. Another very interesting consequence of our approach is that it clarifies the manifold nature of the "locus" of diversification of homologous systems of computation, namely:

Loci of computational diversification (preview)

1. The degree of computational complexity evolutionarily attained in each particular case—as in the contrast between language and other primate's alarm systems;
2. The connections evolutionarily established between the Computational System and other "external" (or "performance") systems—as in the contrast between language and birds' knotting activities; or
3. Combinations of (1) and (2) as in the contrasting of language and birdsong.

Inspired by a previous proposal from Camps and Uriagereka (2006), we conclude this chapter by suggesting that the model can be fruitfully applied trying a fresh approach to the Neanderthal question, in that it offers the opportunity to examine and interpret from a cognitive and linguistic point of view the contrast between the archeological record associated with this species and that of anatomically modern humans. This is another welcome result of our model, given the failure of anatomic and, more recently, genetic approaches to clarify the question (Balari et al. 2011, for a review).

We close the book with a brief chapter of conclusions (Chapter 8: Conclusions), in which we also advance some intriguing avenues of research opened by our architectural model of the mind. We have also added an appendix (Appendix: "On Complexity Issues") on mathematical questions that may be useful for those wishing to go deeper into the details of our proposals concerning computational complexity.

The more biologically oriented chapters of this book will therefore be devoted to progressively constructing a picture in which another crucial element of our argument throughout this work will stand out, namely that there is something else about language that Wittgenstein, and many others before and after him, failed to see: Language is an anomaly—a monstrosity we dare say, both culturally (as an invention) and organically (as a part of our body); and it is this anomalous, monstrous, character of language that we believe justifies giving epistemological priority to its organic aspect: The cultural anomaly some like to call "language" could never, ever, have been invented without there first being some organic monstrosity that made it possible in the first place. A monstrosity, we will argue—adhering to the ideas and assumptions of the teratologists of the nineteenth century, that strictly obeys the organizational principles of the organ in question.

This eventually takes us to the next chapter of our book (Chapter 2: "My Beloved Monster"), whose presentation we decided to leave as a coda to these introductory remarks, as it is our "just-so story" in which we relate how humans endowed with an anomalous CCC could survive and ultimately proliferate in a context in which they most probably looked like "monsters" to conspecifics. No single biological feature is advantageous in itself: It depends on environmental and populational contingencies. CCC_{HUMAN} is no exception to this elementary principle. But in this chapter we move one step further and argue that the ultimate evolutionary success of CCC_{HUMAN} was crucially dependent on its original disadvantageous character in a populational context in which such a strange ability was seen as a cause of exclusion and stigmatization. We depict this process as a case of "sympatric speciation" (Mayr 1963), with a group of people (a "founding population") preserving the feature within the larger group precisely because the feature acted as a signal preventing interbreeding. We contend that the founding group proliferated when the potentials (or "adaptability;" Reid 2007) of CCC_{HUMAN} showed up in a dramatic environmental and demographic crisis, of which there appear to exist evidences coming from populational genetics (Behar et al. 2008). Within our overall model, it is important for us to explain very meticulously that nothing of what we say in this chapter "explains" the emergence of CCC_{HUMAN}: If our ideas are on the right track, they simply explain its proliferation. Thus, from an evolutionary perspective, the emergence of CCC$_{HUMAN}$ is an accidental outcome of an ancient system of development, while its proliferation is an accidental outcome of very particular environmental conditions immediately following its emergence. Explanation cannot work the other way around, contrary to what is routinely assumed within the communicative approach, actually a branch of the adaptationist paradigm. Thus, we are in complete agreement with one of Fodor and Piattelli-Palmarini's (2010) main lessons: Once we exit the internal mechanisms of individual development and reach the individual as an agent for complex environmental interactions, we abandon the domain of theory and enter the realm of history.

So what we are about to tell you here—dear reader—is the story of a monster who strived to survive...and won. This book is assumedly assembled using large doses of theoretical imagination and a light touch of historical (science?) fiction. Or so we tried. Readers will judge.

2

My Beloved Monster

> Thus a species that possesses linguistic competence may indeed take over the earth as a consequence of the technological and managerial capabilities that are the result of language, but in a species lacking linguistic competence, the rudimentary ability to form linguistic elements by a few individuals may be taken as a sign of difference that causes them to be expelled or even killed.
>
> Richard Lewontin, 1998

This chapter aims to clarify some preliminary questions on which the argument that we unfold during the course of this book deeply relies. First, we introduce the idea that the Faculty of Language is the outcome of a process of evolutionary normalization of a developmental abnormality—or "monstrosity," and then we outline how such a process could be brought about through the action of ordinary biological processes. The first section of the chapter focuses on the environmental/populational side of this question, while the second explores more organism-internal aspects of the matter.

Before proceeding, it is worth keeping in mind that the expression "Faculty of Language" is intended here to refer to a complex biological structure, comprising a natural system of computation and a number of connections with other peripheral or external systems. We are aware that in doing so we depart from other standard uses of the same expression, where it rather refers to an innate system of language-specific representations and mechanisms. While we are not denying that such a system actually exists in the mind of humans, it is crucial for readers to understand that we will be adopting a much more basic level of analysis in this book, at which—we believe—the ultimate biological requirements in order for linguistic representations and mechanisms to be of any use to a human mind are to be found. One of the main contentions of this book is that the biological structure that we have in mind—a natural system of computation + external connections—is not human specific; however, particular properties of its human variant—regarding both computational power and external connectivity—are the necessary conditions for humans to create, acquire, and use systems of linguistic conventions.

But this is not a book about how conventional linguistic systems started to exist: Our more modest goal is to explain how the basic biological requisites for this to be possible emerged at a certain point in human evolution. So it should not come as a surprise if we affirm—as we actually do throughout the chapter—that the existence of the Faculty of Language predates the existence of languages, and that it either emerged with no particularly associated function whatsoever or it emerged with associated functions not even remotely resembling language—an idea that is very likely to be generalized to any biological system and its putative functions.

A terminological twist can perhaps be helpful to make the idea more palatable to readers, so imagine that we decide to use the term Central Computational Complex (CCC) instead of Faculty of Language (FL), to refer to the above-mentioned biological structure—i.e, a natural system of computation + external connections. An advantage of this term is that it is readily applicable to the same structure in different species, with each variant exhibiting a particular level of computational power and a different pattern of external connectivity. Besides, we think that the idea that "the existence of a CCC with human-specific properties predates the existence of languages, and that such a CCC either emerged with no particularly associated function whatsoever or it emerged with associated functions not even remotely resembling language," loses much of the controversial air that it has when formulated in terms of FL as in the previous paragraph—both formulations being however fully equivalent for us.

Whatever we decide to call it—not an important issue for the particular purposes of this chapter, it is this system that we contend that emerged as a developmental monstrosity among early humans, and ended up as a normally stabilized species-specific character mostly due to its historically acquired linguistic applications. Let's see how.

2.1 The Monster Untamed

We do not know when humans began to speak—or, as we prefer to say, we do not know when FL appeared as a component of the human mind; not exactly the same thing, as we will explain. Actually, nobody knows for sure. Some hints of innovative forms of behavior—that paleoanthropologists customarily connect to the presence of linguistic abilities (d'Errico et al. 2003)—are maybe as old as 300 ka,[1] but the whole "modern cognitive toolkit" of humans was most probably not fully in place until 100 ka ago (McBrearty and Brooks 2000). It is impossible to offer a convincing interpretation of this crucial time span as to whether FL was there from the very beginning or

---

[1] In the symbolic domain, the processing of pigments has been dated at $c.280$ ka (Barham 2002); in the technological domain, innovative points and blades have been dated at $c.250$ ka (McBrearty and Brooks 2000).

not—i.e., it rather appeared at some other point along the road leading to the ultimate forms of modern cognition/behavior. It is our conviction that many of the earliest behavioral novelties that have been connected to the existence of language in the literature—especially the emergence of symbolic cultures—are definitely uninformative in this respect, as they can conceivably exist in the absence of FL—see Balari et al. (2011) for a complete argument. On the contrary, certain technical abilities—the first hints of which are some relatively late manufactured ornaments,[2] but that most probably were also important in other domains, such as clothing or housing, and from an earlier time—reveal underlying cognitive processes of a complexity comparable to that of processing linguistic expressions (Piattelli-Palmarini and Uriagereka 2005; Camps and Uriagereka 2006). We believe that they are the most reliable evidence so far pointing to the existence of FL at some particular point of human evolution. It is not clear, however, whether Neanderthals were capable or not of such practices, which means that we do not know for sure at exactly what point this cognitive endowment was already in place—did it evolve at such an early time as to include Neanderthals or was it rather a late evolutionary outcome exclusively affecting Anatomically Modern Humans? The question is extremely obscure: Recent findings in the field of paleogenetics seem to favor the former conclusion (Krause et al. 2007), while the archeological record continues to be much closer to the latter (Mellars 1998; d'Errico et al. 2003; Balari et al. 2008). We return to this topic at the end of this book—Chapter 7. For the time being, let us conclude simply by saying that little more can be said in this area that is not speculation.

So let's speculate.

The place is Africa—we do not know exactly where, some 270–80 ka ago. The actors are members of a population of humans—maybe transitional archaic humans, maybe anatomically modern ones (Krause et al. 2007; Klein 2009). Something weird is happening to these people—well, to just a few of them. A certain brain structure—let us call it X, for the time being—has undergone an abnormal overdevelopment in their heads and has entered into contact with other preexisting brain components. To put it plainly: They are monsters—even if they do not look like them. Of course, they do not know that they are monsters—and neither do people around them. However, they do bizarre things and—what is worse—they cannot refrain from doing them. This is the bad thing about being a monster: You cannot behave except as a monster, so everybody else starts rejecting you. Fortunately, you are not the only monster in town. Monsters never walk alone.

But what does being a monster mean exactly?

In a way, *monster* is a word almost devoid of meaning, as almost everyone is a monster from some other point of view. It is just another member of a family of

---

[2] We are referring to the marine shells perforated for ornamental use found in Blombos (South Africa) and dated at 70 ka (d'Errico et al. 2005). We return to them in Chapter 7.

words—with *madness, pathology,* or *cannibalism* being other conspicuous members (Foucault 1961; Canguilhem 1966; Cardín 1995)—that have been traditionally used to refer to those who are different from us. However, every word has many lives and, in some of its lives, *monster* is a word with very clear and precise meanings. We are interested here in two of them.

The first one is the technical sense that the word had in the writings of prominent anatomists of the nineteenth century—among them, Étienne Geoffroy Saint-Hilaire and his son Isidore, Étienne Serres, John Hunter, and Richard Owen; see Richards (1994) and the papers in part I of Fadini et al. (2001). For them, the study of organic abnormalities was a field of scientific inquiry worth taking very seriously and, as a matter of fact, they transformed the interest and anecdotal registration of such phenomena into a scientific discipline—teratology. Classical teratologists found monstrosities extremely important as primary sources of biological information for three different—but closely related—reasons: (1) They do not occur at random—i.e., not every conceivable teratology is a possible teratology; (2) they are the consequence of abnormal courses of development—opinions diverged as to whether precocious births or elongated prenatal stages were the main source of monstrosities, as well as to whether they were mostly due to environmental or internal causes; and (3) they have evolutionary import—an idea predating in all cases the irruption of Darwinian evolutionism. As for the first question, classical teratologists guessed that monstrosities obey certain regularities and they even formulated some statements with a law-like character, like Geoffroy's "affinité de soi pour soi"—double malformations always take place between homologous parts—or Hunter's law—all supernumerary parts are joined to their similar parts, as a head to a head (figure 2.1). Teratologists of course understood that most monstrosities were deleterious. However, in those exceptional cases in which abnormalities were not harmful, classical teratologists believed that they represented an empirically well-attested mechanism of species transformation. Thus, as monstrosities have an evolutionary potential and obey law-like principles, evolutionary paths are predictable to a certain extent—according to this "teratological evolutionary theory" (Richards 1994); and, as monstrosities are the outcome of unusual paths of development, patterns of evolutionary change can be glimpsed by studying the early growth of individuals—also according to this point of view.

Darwinism cast doubts on the evolutionary significance of teratologies—in Richards' (1994) felicitous expression, "Darwin domesticated the monster"—and the idea that evolutionary outcomes are the accumulative effect of minute changes over long periods of time has become orthodoxy now. However, biologists like William Bateson, Hugo de Vries, Richard Goldschmidt, and Pere Alberch kept scientific interest in monstrosities alive—even if marginally—within the evolutionary thought of the twentieth century. Leaving aside particular interests, personal influences, and the state of the art of biological disciplines at different stages of the

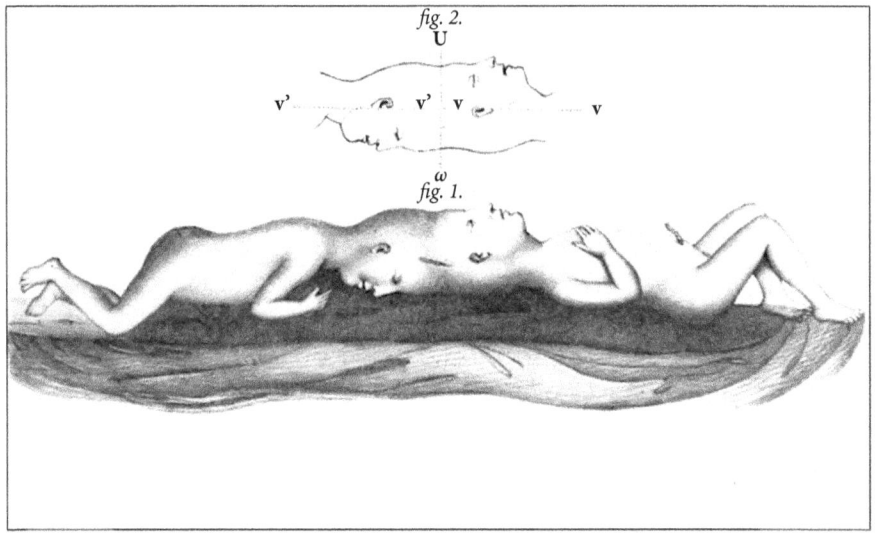

Martin fecit. Lith de Delaunois.
Fig.1 et 2. Céphaliade.

FIGURE 2.1 "... El cirujano no ha querido | finalizar lo comenzado..." ("... The surgeon did not want | to finish what he started...")
Monstrosities are exceptional outcomes of development that do not occur at random. There is a "logic" underlying them (Alberch 1989), the understanding of which provides bona fide information regarding the evolutionary potential of organisms.

_{In the image: a cephalopagous conjoined twin, from I. Geoffroy Saint-Hilaire (1837: Pl. XIX; Fig. 1 and 2). Caption from the song "Unidos" ("Conjoined"), by Parálisis Permanente (1982).}

century, some generalizations can be made as regarding these authors' concerns and ideas. One of them was their inclination to favor "saltationist" models of evolution—against the rampant "gradualism" of the Darwinian mainstream; another was a relatively temperate treatment of the practical benefits of organic structures when explaining their evolution—compared to the radical functionalism of Darwinian adaptationism. The following passages from Pere Alberch (1989) epitomize this, while at the same time illustrate the connection between "modern teratologists" and their nineteenth century predecessors:

"Monsters are a good system to study the internal properties of generative rules. They represent forms which lack adaptative function while preserving structural order. An analysis of monsters is a study of pure form." (p. 23) "These major deviations from normal development result in forms that are often lethal, and always significantly less well adapted than their progenitors. Therefore, one expects monsters to be consistently eliminated by selection. This is a useful property because if, in spite of very strong negative selection, teratologies are generated in a discrete and recurrent manner, this order has to be a reflection of the internal properties of developmental systems." (p. 28)

Monsters are not trendy, but for us they are still hopeful. So "hopeful monsters" (Goldschmidt 1940; Theißen 2009) will be the main characters of the story we are about to tell. For the moment the previous paragraphs are a justification of our adherence to the Geoffroyan–Alberchian teratological tradition—technical details will have to wait until Chapter 6.

The second meaning of *monster* that we are also interested in has sociological coloring, as it has to do with the other's view of difference as abnormality. However, our social monstrosities are also anchored in biology—and not simply for their teratological origins, as we will explain presently. This sense of the word would be the "sociobiological" concept of monster, were it not for the specialized use of the term in the second half of the twentieth century (Wilson 1975), as well as the connotations acquired after years of intensive criticism (Lewontin et al. 1984). So we will use the expression "social monster" instead, a notion by the way not entirely alien to anthropological studies and to which Foucault (1999) appealed in his detailed historical study on abnormality when he wrote:

[C]e qui définit le monstre est le fait qu'il est, dans son existence même et dans sa forme, non seulement violation des lois de la société, mais violation des lois de la nature. Il est, sur un double registre, infraction aux lois dans son existence même. Le champ d'apparition du monstre est donc un domaine qu'on peut dire «juridico-biologique». (p. 51)[3]

Note that Foucault also sees social monstrosities as grounded in biological monstrosity, the main difference between our proposals and Foucault's being that we locate the emergence of social monstrosities at a time antedating the apparition of fully fledged legal institutions. Thus, our contention is that, before becoming "hopeful," our monsters first had to become "social."

How could this have happened?

Our monsters are cognitive monsters. An abnormal brain structure X is in their heads so they do bizarre things. We cannot say exactly what they do at this early stage of the story. It is easier to say what they do not do: They certainly do not speak like we do today. To do so, grammatical traditions have to develop, but these require the existence of speaking communities, which do not yet exist. At this point, the only things that do exist are a few monsters who do strange things. They probably move differently compared to the rest of the people, they manipulate things in ways unlike anybody else and the sounds they make sound different. They act as if what they did lasted longer and was more carefully done—as if they were more consciously dedicated to ongoing patterns of action rather than their practical benefits. This is

---

[3] "[W]hat defines the monster is the fact that its existence and form is not only a violation of the laws of society but also a violation of the laws of nature. Its very existence is a breach of the law at both levels. The field in which the monster appears can thus be called a 'juridico-biological' domain." (Translation by Graham Burchell.)

not the way conspecifics see them; "normal" people simply see them as the people who do strange, useless things. So the conditions are set for these cognitive monsters to be stigmatized, marginalized and in the teratological sense to become social monsters too. Curiously enough, it is at this point that the social monstrosity is transformed into a new biological condition, which in time will have drastic consequences. Let's see why.

Imagine that we have a "single, initially randomly mating population" and that, for some reason, a "reproductive barrier" comes up within it "without any spatial segregation" (Futuyma 2005: 380, 393). This would be an incipient process of "sympatric speciation" (see Mayr 1942, 1963; Futuyma 2005: Ch. 16, for an introduction). Barriers to gene flow can be of very different types, ranging from geographical isolation to gametic incompatibility—see Coyne and Orr (2004); Futuyma (2005: 15; Table 15.2), for a summary. One such mechanism is known as "behavioral," "sexual", or "ethological isolation," meaning that the recognition of differences in the way individuals act may have the effect of preventing them from mating. Indeed, there exists a concept of species called the "recognition concept of species" (Paterson 1985)—where organisms belong to a same species to the extent that they recognize themselves as such, otherwise they will avoid mating. We do not take our imaginary tale that far. However, we think that individuals carrying X in their heads and behaving in maybe subtly unusual, but nonetheless bizarre ways, were strange enough as to be recognized as different and condemned to sexual isolation. This does not prevent them from mating with each other, so a relatively small group of stigmatized, maladapted people would have established itself as a part of this human population (figure 2.2).

Our story is not over yet, but some critical observations can be made. According to our tale a small, differentiated group has appeared within this human population, but for no reasons that have anything to do with the advantages associated with their distinguishing feature. It is actually these people's ineptitude in carrying out normal human practices that has placed them in a marginal, but stabilized corner of humanity. So, this is not exactly—or not properly speaking—a case of speciation, but rather something closer to what Gregory Bateson (1949) described as

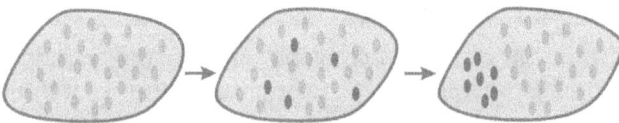

FIGURE 2.2 "One day the world will be ready for you and wonder how they didn't see." Behavioral isolation potentially leads to barriers to gene flow and founder effects within formerly single mating populations.

Caption from the credits of the album *Beautiful Freak*, by Eels (1996).

"complementary schismogenesis," meaning that the more or less markedly different behaviors of two social groups may elicit their segregation, but in lasting terms of contact and equilibrium.[4] However, when this pattern also functions—as we speculate—as a reproductive barrier, it acquires—beyond its sociological character—a biological import.

People with X in their heads are indeed maladapted people within this scenario—they are misfits. The only reason for their genes to become limitedly fixed in the pool of this population is that they have somehow unexpectedly found a sociological niche in which to drift. This does not mean to say that these people's cognitive phenotype is irredeemably disadvantageous: Far from it—as we well know—it is full of possibilities, waiting for the right conditions to express themselves. So X—whatever X happens to be; and you will have to be patient until Chapter 5—is not an adaptation: It did not evolve because it was apt for its function right from the beginning; it appeared as a developmental accident and not a particularly advantageous one. X is, however, a perfect example of Reid's (2007) concept of "adaptability"—i.e., "a feature [that] may not be seen to be 'for' anything in the short term" (p. 13), but may help the organism "to adjust itself effectively in response to internal and external environmental change" (p. 141). Note that "adaptability" is not a form of adaptation with some mysterious "look-ahead" capacity. Adaptability is something different: A property of organic forms, which describes their capacity to behave differently where their internal or external environments are also different—see the next section for further clarifications.

So the question is: When and why did the state of things change as to allow X to show its full evolutionary potential? We are near the end of our tale.

The biological trick that we need in order to transform our monsters into hopeful monsters has a name: Bottleneck. For some reason, the number of individuals of a population dramatically decreases, with the effect that the genetic makeup of the few who survive will spread into the population at large if the new colony persists and grows after the crisis (Futuyma 2005: 232). It was in a scenario like this with a critical population decline under extreme environmental conditions that we think that X proved extremely advantageous for its bearers. So survivors were mostly monsters or, rather, former monsters transformed into a "founder population" in which X was but a normal feature (figure 2.3).

If the bottleneck story is the *deus ex machina* of our evolutionary tale then the punch line where most people happen to have X in their heads is its happy end. Granted—but...

---

[4] Otherwise, the small population of monsters would never survive. This is why we prefer a model of sympatric isolation over an allopatric one, in which case monsters would be expelled and—as competitors of the much bigger predominant population—rapidly killed. So regarding this point we somehow depart from Lewontin's otherwise insightful remark, with which we open this chapter.

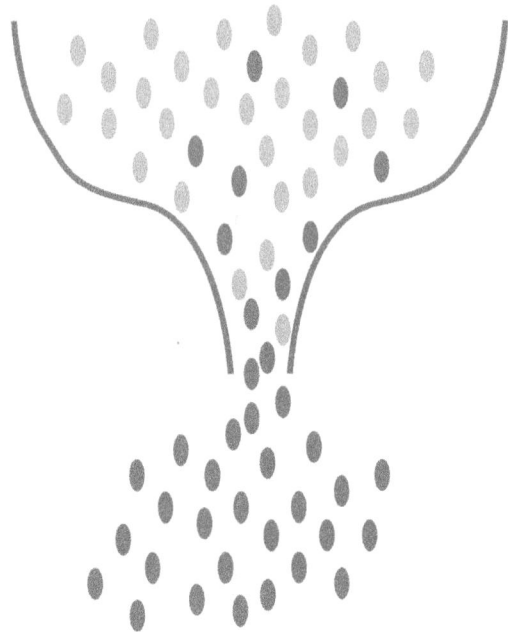

FIGURE 2.3 "Delta Sun Bottleneck Stomp"
New environmental conditions potentially lead to population crisis, where formerly exceptional phenotypes may become the norm.
Caption: song title from the album *Deserter's Songs*, by Mercury Rev (1998).

Bottlenecks in human population have been repeatedly proposed at around the time our story happened and independently of the question on the origins of language. In some cases, meticulous details concerning the time, place, and causes of both bottlenecks and the releases have been proposed. One such proposal is that of Ambrose (1998), according to whom the eruption of Toba (Sumatra)—the largest known explosive eruption of the Quaternary, dated at 71 ka—was followed by six years of volcanic winter and the coldest, driest millennium of the late Pleistocene, which "may have caused low primary productivity and famine, and this may have had a substantial impact on human populations" (Ambrose 1998: 635; see also Rampino and Ambrose 2000, and Gathorne-Hardy and Harcourt-Smith 2003, for some critical comments). The release from this bottleneck, also according to Ambrose (1998), could have been as late as 60 ka, and he connects it—following Harpending et al. (1993)—with the emergence of innovative technologies in sub-Saharan Africa, where tropical refugia offered the best conditions for surviving. Hetherington and Reid (2010: Ch. 7) suggest that the Toba eruption just contributed to make things worse, since climatic conditions were already quite harsh, as the Earth had entered Marine Isotope Stage 4 (MIS4) corresponding to what is known as the lower Pleniglacial. Note, moreover, that MIS4 lasted until 60 ka ago, when MIS3—described as cool, dry, and highly unstable by Hetherington and Reid (2010)—started, in coincidence with Ambrose's (1998) dating for the release of the bottleneck. In fact, Hetherington and Reid (2010: 223ff.) also support the idea that this period was critical for the expansion of novel cognitive abilities.

Alternative causes of the bottleneck/release effect are also conceivable—epidemic diseases, global climate change, dispersing across land bridges or narrow straits; see Lahr and Foley (1998). Regardless of which solution may be correct, what seems likely is that something akin is at the base of the present pattern of diversity exhibited by human populations—see Fay and Wu (1999) and Behar et al. (2008); Hawks et al. (2000) for a critical view.[5] It also seems to be at the base of the present pattern of diversity exhibited by phonemic inventories across languages, according to Atkinson's (2011) comparative study of more than 500 languages,[6] which supports the idea that:

> A cultural founder effect operated during our colonization of the globe, potentially limiting the size and cultural complexity of societies at the vanguard of the human expansion. An origin of modern languages predating the African exodus 50,000 to 70,000 years ago puts complex language alongside the earliest archaeological evidence of symbolic culture in Africa 80,000 to 160,000 years ago. (Atkinson 2011: 348)

Note that Atkinson treats "phonemic inventories" as cultural conventions, as we do with grammar at large in this book. So Atkinson's conclusions are fully compatible with our idea of a relatively early emergence of FL, followed by the release of language-related cultural practices after a population bottleneck.

It is time to repeat it again: We don't know whether—or to what extent—our story is true. We don't even know how exactly we could test it—even whether it is somehow testable, to begin with. However, the good news is that we have a very detailed idea—and, we think, a testable one—about the main character of the story: X—what it is, how it evolved, and how we put it to use. And, best of all, what we know about X is fully compatible with our just-so-story: It is a "monstrosity," it did not evolve "for" nothing, and it is "adaptable" enough to be useful in the most varied domains, from motor control to reasoning—especially if people around you don't think that your behavior is weird.

But, before introducing you to X, we still need to explain something about how it could become an advantage after being an anomaly.

---

[5] To be fair, approaches as to the origins and diversification of modern humans based on this idea actually refer to a number of more or less parallel "bottlenecks," rather than just one. Nothing in our "just-so-story," however, is incompatible with this scenario. Our story is, apart from being highly speculative, very schematic. But within a scenario of global crisis due to catastrophic causes, we think that the same effect could be repeated at different focuses, followed by the dispersal—in some cases maybe the grouping—of survivors after the release.

[6] But see Hunley et al. (2012) for discussion.

2.2 The Adaptable and the Plastic

Monsters with X in their heads had linguistic capabilities but did not speak. It is our conjecture that "heads with X" only became "talking heads" later on, some time after the population release that transformed them into ordinary people with X in their heads. We think that they were capable of putting their linguistic abilities before properly speaking to good use and that this was the key to their evolutionary success. How was it possible for creatures that did not speak to be however capable of striving due to their linguistic talent? To solve this paradox, we need to understand the following statements:

1. Humans did not start speaking until they developed grammatical traditions—maybe along the lines of Jackendoff's (2002) incremental model, as we are simply committed to a saltationist view concerning the biological requirements for creating grammars, not grammars themselves;
2. The organic system that allowed them to develop grammatical traditions was there before they started doing so—as a consequence, we argue, of an abrupt cognitive change;
3. Humans were capable of putting the system in question into use for a number of different purposes before they started using it for creating and exercising grammatical conventions.

In the introduction to this book we explained the difference between FL and grammars, and we have also advanced in this chapter the idea that the former is but the human-specific version of a CCC underlying the creation, acquisition, knowledge, and use of the latter. This should be enough in order to understand points 1 and 2 above, which simply state that FL—an evolved organic system—was there before humans invented grammars—which are cultural conventions. It was only at this latter point that humans began to speak—and speaking communities came into existence. Such an idea should not come as a surprise, as FL is not—according to our theoretical framework—an adaptation "for" anything in particular, including speaking. We think humans were capable of benefiting from FL before they "discovered" its benefits as a means for speaking fully fledged—i.e., grammatical—languages. Furthermore, for any of its putative original uses—whatever they happened to be, FL had to have already been there before humans started to use it in the corresponding areas—in other words, FL is not an adaptation. Such a statement would look tautological, were it not for the fact that it reverses the logic of extreme yet popular forms of adaptationism, according to which given the pressure to perform a function and enough time, any organism—be it an ant or a hominid (Bickerton 2009)—will evolve the system to uphold it.

We prefer to position ourselves on the opposite side of the spectrum, defending the argument that functions only start to define themselves once the system that upholds them is already in place, connected to a more or less wide range of other organic systems and to certain environmental conditions. We deny the role that functions are customarily endowed with in evolutionary explanations to the extent that we think that the adaptationist paradigm relies on a highly dubious contention: Namely, that functions somehow preexist their biological incarnations—and, even more dubiously, that these "disembodied" functions exercise a causal role in evolution. For us, functionalism is thus a form of Platonism—to be avoided if a better explanation is available. And we think that a better one is definitely available.

According to our own point of view, organic systems have a "functional potential"—rather than "functions," the limits of which are structurally given—i.e., it is the form of the system that determines its functions, and not the other way around. This is not to affirm that such a potential is fully manifest at every point in the history of the said structure. We believe that it varies from one time to another—certain functions access the foreground, while others remain in the background or are even dormant, depending on the structure's connections to other organic structures and to the environment. In other words we could say that functions are indistinguishable from the structure that hosts them; either when pinpointing a particular function or a collection of functions both proper or typical, thus we are just focusing on the structure from a dynamic point of view and applying it at a particular, contingent point of its historical development.

This idea is fully congenial with the theses put forward by Reid (2007), according to whom functions do not explain the evolutionary success of organic structures. So, the higher the degree of specialization of an organic structure the lower the chances for its bearers to survive in scenarios of rapid and radical change in conditions. For Reid, being adapted to present circumstances is a poor evolutionary strategy; a much more advantageous alternative from an evolutionary point of view is to be endowed with a high degree of "adaptability," which means that the organism is capable of reacting with a more or less open repertoire of answers to changing, unstable environments—or of exploring and colonizing new ones. In Reid's own words:

While adaptations are inflexible, adaptability is the quality of an individual organism to adjust itself effectively in response to internal or external environmental change. (Reid 2007: 141)

Our own interpretation of Reid's ideas does not strictly contemplate organic structures as being either "adaptive" or "adaptable" ones (Balari and Lorenzo 2010a). For us, structures customarily referred to as "adaptations" are just structures with a low degree of "adaptability;" but, in the end, every organic structure is an adaptable one—even if sometimes with a degree close to zero; absolute "zero" being more an ideal rather than an actual possibility. So, organic structures have a functional potential that in many cases is higher than currently thought, probably because descriptions tend to focus on just one prominent function—given the lifestyle of a species in its present environment, and new functions only emerge when the species' environment or lifestyle unexpectedly change. Think of the large beak of the toucan (*Ramphastos toco*), interpreted as a sexually selected ornament or as a refined adaptation for feeding. Tattersall et al. (2009) have recently argued that it is an effective thermoregulatory device (figure 2.4). But none of these functions is *the* function of the toucan's beak; rather, they all enter inside the structure's adaptability circle and, crucially, without exhausting it. So there exist other putative applications that, were the toucan's internal organization or external environment to change, would naturally emerge as normally performed beak activities—for example, its use as a defensive or aggressive device in an overcrowded medium with scarce food resources.

Reid argues that adaptability is obtained as a consequence of either exploring new environmental conditions—so new forms of behavior may directly emerge, or because the organism somehow adjusts its internal organization—so new organic systems may almost directly emerge:

Emergences are innovations that in their most radical form constitute new levels of organized complexity. Their novelty comes from novel relationships among preexisting systems, combined perhaps with contingent catalytic additions—in lay terms they produce wholes that are greater than the sum of their parts. (Reid 2007: 82)

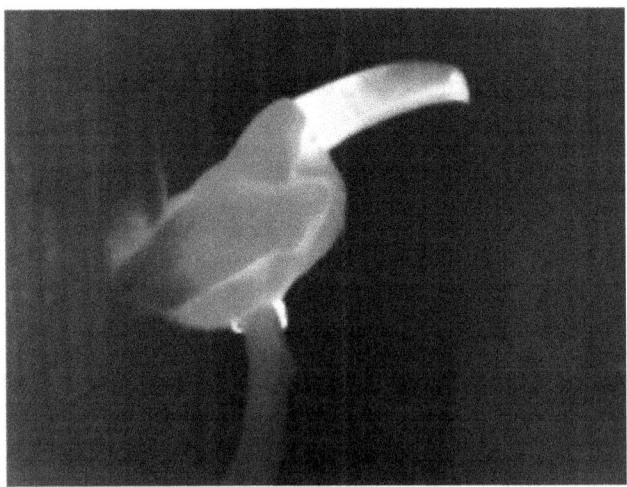

FIGURE 2.4 "Of Moons, Birds & Monsters"
The toucan's beak has recently been described as an effective vascular thermal regulator, as witnessed by this infrared thermal imaging.

Original thermal image from Glenn Tattersall's personal collection and used with permission from Prof. Tattersall. Caption: song title from the album *Oracular Spectacular*, by MGMT (2007).

Combinations of both adaptability conduits are a third, non-negligible source of evolutionary novelties. This is the image that, in our opinion, best fits the evolutionary emergence of FL. For us, FL is an evolutionary emergence in the sense that: (i) It resulted from putting together a number of preexisting cognitive systems, among them, the still mysterious structure X; and (ii) it directly endowed humans with a notably enriched behavioral repertoire, which manifested itself as conditions allowed. These observations are very important because they may help to understand that FL—as in every other case of natural emergence—is clearly rooted in a basic property of organisms—which means that biological emergences are not mysterious at all.

The property in question is "plasticity," defined by West-Eberhard (2003: 34) as "the ability of an organism to react to an environmental input with a change in form, state, movement, or rate of activity." Note that the definition is wide enough to connect the property to environmental stimuli or to radical changes in the species' typical ecosystem, and relate it to organic adjustments from transient behavioral answers to phenotypic reorganizations. Plasticity, according to West-Eberhard, is active at every level of biological description—from molecules to behavior—and at every point of the development of organisms. "Development" and "plasticity" are quasi-synonyms in West-Eberhard's framework—and the expression "developmental plasticity" she uses is almost redundant, since she defines the former as change in responsive or reactive phenotypes due to both internal and external inputs (West-Eberhard 2003: Ch. 5). So, what development/plasticity names within this framework is the life-lasting capacity of organisms to adjust at every level of organization in response to ever changing environmental conditions. It is not difficult to understand the close connection between West-Eberhard's development/plasticity and Reid's emergence/adaptability.

West-Eberhard points out that an underlying source of developmental plasticity is the modular organization of biological structures and, we would like to stress, the corresponding factorial or combinatorial character of development (Minelli 2003: Ch. 10; West-Eberhard 2003: Ch. 4). Modularity helps the reorganization of organic designs at least relatively to conceivably more rigid patterns of organization. So, for example, a collection of independent or poorly connected brain systems is a probable source for a biological emergence, which will possibly multiply the effects of isolated evolutionary transformations of its component pieces. Not every single case of connectivity so attained in development will acquire evolutionary import. For this to happen two further conditions seem to be crucial: (i) That the relevant connections prove advantageous—once the organism has had the opportunity of "experimenting with" them (Reid 2007); and (ii) that they undergo "phenotypic accommodation." Both points underline the idea that, before they operate, biological emergences are "anomalies due to plasticity" (West-Eberhard 2003: 83). As for the second point, many mechanisms for accommodation to occur seem to exist—regulatory mechanisms at the genetic level being a particularly active one. The following chapters will illustrate that all these elements fit in perfectly well with our own image of the evolution of FL.

Plasticity does not stop at the point at which an organic structure is already in place. A more reasonable assumption is that, with the new structure, something like a "functional release" occurs and new ways to act come into existence. This does not mean that new structures arise with an associated function, or a collection of associated functions; it just means that new structures pave the way for organisms to react in new, innovative ways to the pressures of the environment, which is, according to West-Eberhard, just another facet of plasticity.

The behavioral potential of a new structure is fully given from the very moment it emerges—which is why we affirm that functions are indistinguishable from the structures that exercise them. However, the largest part of this potential remains dormant in most individuals most of the time, as it only comes to light if some particular environmental input triggers behavioral reactions located within such a potential. This also means that, in order for behaviors to come up with something like a typical habitual repertoire two conditions seem to be required, as in the case of structures themselves. The relevant inputs are expected to be sufficiently recurrent to be taken as a part of the normal environment of the species—or of the "niche" it inhabits and constructs (Lewontin 1983; Odling-Smee et al. 2003); and the behaviors in question must undergo something akin to "phenotypic" and "behavioral accommodation." So for people to start properly speaking, they needed to be already endowed with FL, which was probably active in other areas still in the technical domain. This was a necessary, but on its own insufficient condition: Grammatical conventions still had to develop to accommodate the primeval forms of vocal behavior leading to language as a typical component of the behavioral repertoire of humans. We pointed out before that grammars are apt to perform this function as external devices that serve to scaffold the learning process. However, it is probable that they also fulfill this role of behavioral accommodation by creating the conditions of uniformity among individuals required to support communities in which every kind of tradition can flourish, be held, and transmitted—grammatical traditions included. Grammars have, from their very character, a normative import that has very likely historically benefited the sociocultural development of humans in different ways.

This book is not about "language-as-an-invention"—or grammar—or "language-as-a-behavior"—or speech, because we are convinced that when we enter into these realms, "language-as-a-faculty"—or FL—starts to fade away. Speech—like any other form of behavior—is transient, and would be an amorphous or, at best, a very irregularly organized phenomenon, were it not for its sociohistorical underpinnings—i.e., grammars. These are questions that we examine at length in Chapters 3 and 4, where we conclude that categories coming from the analysis of behavior can only misrepresent the description of language as an evolved phenomenon. So we are going to concentrate on FL instead, which means focusing our attention on the organic structure that made speech-related behaviors possible to begin with and—besides—paved the way for the historical development of grammars and the establishment of traditional speech communities.

We think that this is an important move—which is not to say that we consider efforts directed to clarify the growth of language in these other sociohistorical dimensions unimportant. We believe that languages could never have appeared were it not for the fact that humans were already endowed to create them. So we consider it urgent to break with the idea that the emergence of language was just a question of time once humans had reached the right environment—which imposed on them some crucial necessity—or they had started to construct the right niche—as would have also been the case, as conventional wisdom goes, with any other species that happened to occupy similar environments and suffer from similar needs—see Balari and Lorenzo (2010b, 2010c), for some criticism. In the following chapters, efforts will be directed to offer a biologically informed alternative to this naive view on language evolution.

3

The Dead End of Communication

> There is no study of 'language' ranging from ants, to chimps, to human language, to formal arithmetic, any more than there is a study of 'locomotion' ranging from amoeba to eagle to science-fiction space ship; or 'communication,' ranging from cellular interaction to Shakespeare's sonnets to 'intelligent' extraterrestrials.
>
> Noam Chomsky, 1993

An important part of the two preceding chapters was devoted to unraveling the dense fabric of notions and concepts that often underlie the reference of the word "language" in order to show that this is a complex and multidimensional phenomenon whose study can only be approached by a divide-and-conquer strategy. Our main conclusion was that one must at least identify two different dimensions of "language," namely a strictly organic one—which we identified with term Central Computational Complex (CCC)—and a mostly cultural one—for which we reserved the term grammar. This distinction was motivated on evolutionary grounds, for, as we argued in Chapter 2, we believe that the emergence of "language" is better conceptualized as the result of two temporally connected events, corresponding respectively to the emergence of the organic CCC followed by the emergence of grammars. Note, however, that although we see the emergence of the organic component as a necessary prerequisite for the eventual development of grammatical systems, we do not believe that there is any causal connection between the two, in the sense that we do not conceive of grammars as the necessary outcome of the emergence of a Computational Complex of a specific sort. Thus, just as the evolution of the Computational Complex was the result of the fortuitous interaction of a number of developmental, genetic, and environmental factors, the evolution of grammars was made possible through the serendipitous interaction of a number of cultural, social, environmental, and biological factors, with the presence of the Computational Complex being just one of them.

Be that as it may, we have already stressed that the main goal of this book is to articulate a proposal for the evolution of the biological prerequisite for grammars, namely the CCC, and that we have little to say about the emergence of grammars beyond the sketchy remarks we made in the previous chapters. Methodologically, we believe that this is a very promising approach: First, by identifying two different entities that are often lumped together in most uses of the word "language," it has some important and interesting consequences for the evolutionary study of language and animal cognition in general, some of which are explored in the final chapters of this book; second, it avoids what from our point of view are a number of problems and shortcomings of most traditional approaches to the evolution of language. In this chapter and the following we try to identify these problems and shortcomings before proceeding to the presentation of our own proposals in Chapters 5 and 6.

The perspectives we will adopt in this chapter and in Chapter 4 are slightly different from each other. Thus, although both chapters may be read as variations on the same theme, namely the debate between functionalism and structuralism, this one approaches the problem from a more philosophical perspective, whereas the following adopts a more biological stance. Both, however, try to argue in favor of the primacy of form over function, and in both the notion of "communication" plays an important role, to which we turn immediately.

3.1 The Received View

1. Communication is an evolved organic function to the same extent that, for example, blood circulation and locomotion are evolved organic functions.
2. Language, the communication system specifically used by human beings, is therefore a particular instance resulting from the evolution of the communicative function.

These two statements, commonly seen as unproblematic, underpin the foundations on which the two most generally accepted explanations for the evolutionary origins of language are constructed. According to the first, language is a modified descendant of some other ancestral form of the communicative function: That is, it is a new form of communication derived from a preexisting form of the same function, on whose identification the success or failure of the explanation greatly depends. According to the second model, however, language is an entirely new form of communication in the qualitative sense, and must therefore be seen as one of the key innovations introduced during the natural evolution of that function. As a consequence, an evolutionary explanation need not establish specific kinship relations with other forms of the communicative function; rather, the explanation should establish the kinds of (ecological) factors that favored the emergence of this novel way of performing such a function.

As we advanced in the introductory Chapter 1 and as we argued in Chapter 2, these approaches tend to overemphasize the causal role of external factors in evolutionary processes—and, hence, to obscure their inherent dynamic and interactive nature—by either assuming that the term "language" just refers to a behavior or contending that functional/behavioral considerations are prior to other, non-functional ones. Our main goal in this chapter is to try to show that behavioral/functional terms suffer from an inherent lack of specificity and precision that questions their role in the natural sciences. In Chapter 4 we will broaden the context of our argument with a critical analysis of the notion of "behavioral homology" that has found some support in the biological sciences.

The idea that language is not a communication system, one of the main lines of attack to confute these approaches, is not new. We argue, however, that *none* of the so-called "animal communication systems" is really an animal communication system. The notion "animal communication," while perfectly intelligible and even useful for non-theoretical purposes, does not refer to anything that may legitimately be considered a true natural kind. Hence, it is not a notion on which any theoretical approach to language can be exclusively based, nor through which we can explain the evolutionary origins of language. We shall use the term "communicative fallacy" to designate the collection of assumptions expressed in the theories summarized in the first paragraph. The first goal of this chapter, then, is to unburden the evolutionary study of language from the communicative fallacy.

But this chapter also has a second goal. Just as we shall first argue that language is not an evolved version of the communicative function, thereafter we will also argue that language, like any other organic system, is neither an evolved form nor an evolutionary, innovative way of performing any specific function. The assumption that language, or any other organic system, is an object of this kind will be classified as an instance of the "functionalist fallacy." A first step to unveil this fallacy consists in applying the very same argument we have applied to the case of the communicative function to any element f of the set F of organic functions in order to show that none of these varieties actually refer to a natural kind. Evolutionary explanations, we contend, must be based on the identification of homologues of the studied organic systems or of the components of such systems where their complexity calls for such a strategy. Recall that, since the time of its classical formulation, the notion of "homology" has always been insensitive to any kind of functional consideration, and that its more recent reformulations are fundamentally based on the existence of similar developmental resources in the process of establishing the individuality of the implied structures. Only a few contemporary exceptions exist where an attempt is made to extend the notion of homology to functions and/or behaviors, but, as we will argue in Chapter 4, none of these attempts actually succeed in their effort of constructing a consistent definition of "behavioral homology."

We have already argued that an important part of what is habitually referred to by the word "language" must be described as an organic system whose evolutionary explanation needs to be grounded on the establishment of a true base of homologous systems, an idea that we will try to reinforce in the chapters to come, but even adopting, as we will do in this chapter for expository purposes, this more traditional use of the word, we think that "language" is not a modified form of performing any function f; nor in fact is it the modified version of some organic structure s either. Rather, language is the structure in which a developmental system D is expressed, itself a modified variety of an ancestral system which is the origin of many other developmental systems d,[1] each responsible for the implantation in different organisms of different organic structures, both from the functional and the formal point of view.

3.2 Against the Communicative Fallacy

The ability to ward off an environmental threat or to attract a potential sexual partner may be a direct consequence of the ability to receive or emit some kind of signal. It is therefore hard to question the advantages of communication over oblivion. But communication can also be dangerous—one who cries to alert may immediately or directly be exposed to the threat in question; one who trusts a display may eventually find out that it was a fraud or, even worse, a deadly trap. What this boils down to is that minimal differences in the performance of some communicative act may mark the difference between leaving more or less offspring, between surviving or succumbing, depending on a subtle balance between the assumed risk and the derived advantages.[2] In a nutshell, communication appears to be a kind of natural activity perfectly adjusted to the basic principle of *selection between competing variants* on which the Darwinian game of organic evolution is based.

It is thus understandable that communication is easily seen as an evolved organic function with a high adaptive value and therefore as ubiquitous but highly diversified, given the diversity of selective pressures it supposedly had to respond to in different environmental and populational situations (Smith 1977; Hauser 1997). Nor is it surprising that language, the communication system specific to humans, is equally easily seen as a particular form resulting from the evolution of the communicative function (Eibl-Eibesfeldt 1984: Ch. 6). However, we believe that both the idea that communication is an evolved organic function and the idea that language represents

---

[1] Following current genetic conventions, upper vs lower case letters are intended here to represent the human (D) vs non-human (d) versions of the corresponding system (as in, for example, *FOXP2* vs *Foxp2*).

[2] An idea from which, for example, derives the so-called "handicap principle" (Zahavi 1975, 1977; see also Zahavi and Zahavi 1997), according to which this tension favors the evolution of trustworthy signal systems or, in other words, honest forms of communication. Interestingly, the very same considerations have driven some biologists to propose an entirely opposite view from the one fostered by the Zahavis of what communication really is—see Dawkins and Krebs (1978), Dawkins (1982: Ch. 2), and Krebs and Dawkins (1984) for an overview of the idea that communication is not honest but, rather, just a form of selfish manipulation. Be that as it may, this means that not only is communication seen as an expected result of natural evolution, but also forms of communication with very specific design features.

the evolution in a particular direction of this function are unwarranted—and lie at the heart of what we termed the communicative fallacy.

One way to partially challenge the communicative fallacy would consist in arguing that language does not form a natural kind associated with the other varieties of animal communication. Another, much more radical way of calling it into question would be to argue that none of the forms of animal communication actually comprise a natural kind. While the first challenge would be sufficient in order to solve the question as far as human language is concerned, in this section we will complement it with a challenge of the second kind, in order to radically unburden the study of language from the communicative fallacy. As a matter of fact, we will successively develop both challenges as proper parts, of increasing radicalness, of a single long argument, one which we will augment in the next section with an explanation that the communicative fallacy is only a particular case of a much more general and extended fallacy: The "functionalist fallacy." We will then argue that it is in fact the functionalist fallacy that needs to be called into question in order to base the evolutionary study of language on biologically solid and rationally acceptable grounds.

3.2.1 The communicative model: A long history and two living variations on the same fallacy

The communicative fallacy underlies most conceptions, both intuitive and theoretically informed, of language. It is highly difficult to identify some example where one is able to perceive in a clear and unequivocal manner that its influence is not present in some way or another. Moreover, it is certainly present in some of the most influential philosophical approaches to the origins of language anterior to or contemporary with the naturalization of the question by Darwin. For example, Rousseau (1755, 1781), even though he was persuaded that language was a social institution and not a natural instinct, believed nevertheless that language, being the oldest encoding/decoding system and the basis for all other social institutions, should be grounded on some natural instinct for communication, shared with many other animal species, and manifested by the different forms of what he called "natural languages." For Rousseau human "conventional language" was one step ahead of the "natural language" of animals, in that it required some further artificial or social elaboration from the latter. However, he assumed "natural language" to be present at the dawn of human civilization and still alive in vocalizations and gestures that are more directly inspired by our emotions.

It is worth emphasizing that, by means of his reflection, Rousseau introduced two different, yet related, topics concerning language evolution, hotly debated since then: (1) the natural or conventional character of language as a form of communication; and (2) the extent to which language can be said to be a human-specific form of communication. Ideas almost identical to those of Rousseau regarding both questions can be found, for example, in authors like Reid (1764). Others, however, assumed more critical stances towards these assumptions. Herder (1772), for example, denied the existence of any relation between language and the animal expression of the emotions. He also assumed that language was entirely the result of the reflexive capacity of humans put to that specific use, a capacity that Herder certainly regarded as natural, but totally unrelated to other animal capacities. Some time afterwards, Marx and Engels (1845/46) went further to proclaim that not just language, but communication proper (the former being a social instrument for the development and use of the latter) was a unique feature of humans, the only species gifted with the awareness of relationships with the other, for them the hallmark of true communication, whereas, for example, Nietzsche (1887), even though he shared with Marx and Engels the idea that communicative needs played a crucial role in the emergence of language and consciousness, adopted a more naturalistic stance. It is thus clear that the many corners of the debate instigated by Rousseau's reflection meant a close interweaving of the concepts of language and communication, which lasts until today.

In fact, the tendency to associate language with different forms of animal communication is much older than that, although historically it was most of the time alleged to support the exclusiveness of human reason and not as an explanation for the human capacity for linguistic communication.[3] Most thinkers of classical antiquity who addressed the issue took Aristotle's (c.384–322 BC) observations as a point of departure and virtually limited themselves to reproducing them in a similar form. Aristotle classified the different forms of communication according to the three categories of "noise," "voice," and "language." These categories, although intimately related, were useful for him to establish three levels of increasing perfection (a succession of more developed souls or vital complexes) among animals. In essence, a *noise* could be produced with any part of the body—for example, by rubbing air over a membrane under the wall of the diaphragm in cicadas, whereas the *voice* required the activation of the respiratory organ (more concretely, the inspiration of air by the larynx and its concentration and eventual hitting of the trachea), which was made into *language* when it was associated with a specific type of tongue (loose, wide, and lightweight), permitting the articulation of the resulting sounds. On these grounds, Aristotle believed that some birds had language, but considered it absent

---

[3] Lucretius (99–55? BC) is a notable exception (Lucretius 2003: 379–380). He justified the existence of language with the idea that the voice was natural to man just as, for example, song is natural to birds. He appears to have defended the view that it was not possible to conceive of language as the product of some deliberate agreement among the first speakers, and that men spoke with the same naturalness as other animals in the way they behaved and expressed themselves, which made any further explanation of how they understand and are able to understand their fellows entirely unnecessary. Thus, the ideas contained in Lucretius' poem are much more closely related to the debates later opened by Rousseau than to debates concerning language and communication in ancient history.

among viviparous quadrupeds, whose normal means of communicating was voice.[4] He thus defended the view that, among four-limbed viviparous animals, language was a human-specific character.

On the basis of this categorization of animal communication, some authors—Plutarch (AD 45–125) or Claudius Aelianus (*c.* AD 170–249)—came to the conclusion that certain animals possessed, to some extent, a rational soul similar to humans (Plutarch 2005; Claudius Aelianus 1989). However, others—Pliny the Elder (AD 23–79)—considered that, since animals could not deliberately express their emotions, it was only possible to conclude that there was an insurmountable chasm between human nature and the rest of the living world (Pliny the Elder 1998). More recently this is the position we also find among the rationalist philosophers—Descartes (1637), Cordemoy (1668), or Leibniz (1765), who also deemed pertinent to this debate the comparison between language and the other, much more automatized, forms of communication found in animals. Against the rationalist position we find the materialism of La Mettrie (1747), who argued for the continuity of animal and verbal communication, and who considered animals, should the necessary anatomical adjustments be possible, perfectly capable of language. A similar position, along with a brilliant and premonitory presentation of all the details and stances of the debate, was already held by Montaigne (1595).

As for the explanatory models of the origins of language developed after the basic principles of evolutionary theory were established, and which will be the main focus of attention of this chapter, the communicative fallacy also appears as the common ingredient in most of them. In broad terms, one family of models sees in language nothing but a modified variety of an ancestral system of communication. The rival family of models contends that a substantial set of features of language—in fact, the most distinctive or singularizing features of language—might not come from animal communication and have developed *ex novo* during the evolutionary process or, alternatively, might have co-opted features coming from other aspects of animal cognition and behavior. While this set of models may appear to avoid the communicative fallacy, it does not do so precisely because the extension of the evolutionary recipe to non-communicative ingredients pinpoints the particular sense in which language diverges from the communicative behavior of other animals.

The application of the theory of natural selection to the case of language suggested by Darwin in *The Descent of Man* (1871) could be seen as an early, but already quite elaborated, incarnation of the first family of models. Given the context of its inception, we can say that this model is a manifestation of the "Darwinian version" of the communicative fallacy. Darwin, eager to protect the principle of continuity in

---

[4] Aristotle observed that voice was also common among oviparous quadrupeds and dolphins (because they have lungs and trachea), whereas among fish only noise was common. He expounded all these ideas in several fragments of his "biological" writings; see Aristotle (1990, 1999, 2000).

the organic world from the apparent exceptionality observed in many human traits, argued that language, an exemplary case of such singularity, could be put in relation to the quasi-musical calls used by different species in the expression of the emotions associated with the reproductive function (seduction, rivalry, jealousy, consummation, etc.). The similarities between certain aspects of human language and birdsong in many avian species—such as its partially innate and partially learned character or the existence of a phase comparable to human babbling and of intraspecific geographical varieties comparable to dialects—encouraged Darwin to see in birds the true "model organism" for his thesis, as well as the irrefutable proof of the antiquity of the line of continuity thus established. On the other hand, the verification of similar practices among certain primates, especially in gibbons, drove him to the conclusion that the first humans, or their closest ancestors, could also have made a similar use of their voices, originally with a restricted symbolism with sexual connotations, which would have gradually given rise to language as we know it today. Language, in sum, would just be another modified descendant of an ancestral communicative practice, the living testimony of the existence of which would be the different and varied communicative behaviors observed in the animal kingdom.[5] Although the evolutionary explanation of language was taboo in theoretical linguistics for the greater part of the twentieth century, an interesting exception was the case of the Danish linguist Otto Jespersen (1922), who moreover adopted the Darwinian theses concerning the evolutionary relation between the songs of certain animals and language, and the sexual motivation of the most primitive forms of language.[6] Another remarkable exception in the treatment of the problem of language origins in twentieth century linguistics is Hockett (1958), who proposed a model based on the transformations suffered by the inarticulate emissions of the common ancestors of hominids (and, also, of humans) and hominoidea (chimpanzees, gorillas, orangutans, and gibbons), which he referred to as "proto-hominoidea" and of which the so-called "proconsul" he considered to be an example.

This model of evolutionary explanation of language based on the Darwinian version of the communicative fallacy is still kept alive, especially by the work of some primatologists, who also strive to find the connection between the most challenging formal features of human language and the gestures and calls of the other primates. A particularly remarkable proposal in this context is that of Ujhelyi

---

[5] See, in particular, Darwin (1871: 106–114). For a reconstruction of the Darwinian argument and some comments on its (limited) validity, see Lorenzo (2010).

[6] The idea that sexual selection has played a crucial role in language evolution is also rooted in some scenarios proposed within the framework of evolutionary psychology, which, adhering to the idea that communicative acts are better seen as manipulative acts, appeal to the so-called "Scheherazade Effect"—the necessity to develop good communicative skills in order to retain a partner—as one of the determining factors in the process. See Ridley (1993: Ch. 10) for a review of these proposals, and Miller (2000) for a recent formulation.

(1996), who takes over and develops—rather faithfully, but in a seemingly unnoticed way—the ideas originally expressed by Darwin. Ujhelyi contends that the "songs" (or "long calls") uttered in the wild by several species of non-human primates (titis, tamarins, indris, and gibbons), as well as those observed among chimpanzees and bonobos, may well be considered as forms of a "minimal language." Such calls consist in combinations of minor acoustic units, whose level of complexity is fairly variable from one species to the other, with that of the tamarin, for example, consisting in repetitions of two pieces, while that of the gibbon is based on combinations taken from an inventory of thirteen (Mitani and Marler 1989; Geissmann 2000). According to Ujhelyi the songs of these species possess a "minimal syntax",[7] in the sense that the possibility of modifying the lineal disposition of the elements and thus grounding on this the expression of some particular content is (potentially or effectively) already present. Evidently, Ujhelyi acknowledges, it is not a "true" syntax, because true syntax requires special units (inflections or function words) to mark the relations between the combined elements; it is, nevertheless, a stage comparable to certain phases of language development in children or to certain dysphasic disorders, which, for Ujhelyi, is an indication that two different layers of grammar exist dissociated from each other and that the early appearance of the first, simpler layer in language development may be taken as an indication of its early appearance in language evolution too. She observes moreover that in primates these forms of song are only manifested in territorially monogamous species (a very small fraction of them, by the way), such that these different combinations of units serve the function of constructing messages whose content is associated with the social, sexual, and marital condition of the emitter, as well as to its identity. As for non-territorially monogamous species, like chimpanzees and bonobos, the use of these songs, Ujhelyi contends, is an example of the preservation of some ancestral skill in the absence of the environmental and populational conditions that motivated its origin—one more example of the kind of "time lag" explanations offered by sociobiologists and behavioral ecologists for unexpected behaviors from the perspective of strict Darwinian adaptation (Dawkins 1982: Chs. 2–3). Finally, for her, another remarkable feature of this "minimal language" is the "duets" observed in some species between two individuals of a different sex in an interaction that appears to be an incipient form of the bidirectional communication characteristic of human language.

---

[7] In a later paper (Ujhelyi 1998), the distinction (originally made by Marler 1977) is introduced between "phonological syntax" and "lexical syntax" in Ujhelyi's analysis of primate calls, but such qualification does not add any substantial difference in her model. Arnold and Zuberbühler (2006a, 2006b) contend that certain "minimal" forms of "syntax" are also present in the alarm calls of some species of Old World monkeys. They also suggest these might be almost unmodified forms of the kinds of vocalizations from which human language evolved.

Summarizing, although Ujhelyi recognizes that "fundamental differences" exist between the structure and use of human language and any other form of communication, that of non-human primates included, her proposal rests on the assumption that primate communication represents "an intermediate stage between animal communication and language," since it already possesses all the basic features whose further development could have given rise to those properties we perceive as distinctive of human language.[8]

Turning now to the second family of models, these instantiate a much more sophisticated variety of the communicative fallacy, typically observed in approaches which in some way or other adopt a critical stance toward the Darwinian orthodoxy. Perhaps the most illustrative and influential proposal along these lines is the main representative of what we shall call, for reasons to be made clear presently, the "Hauserian version" of the communicative fallacy, which is the one developed by Hauser et al. (2002).

A crucial aspect of the Hauser et al. (2002: 1569) paper is its careful distinction between the "communicative" vs "computational" aspects of language: It makes sense to speak of language qua communication system when we study it from the point of view of its usefulness as a medium to externalize certain kinds of internal mental representations; in addition, however, language must also be studied abstracting away from the mechanisms used for the externalization of these representations *and* from the systems responsible for recruiting their content, in which case language is reduced to a mere computational system in charge of generating "bridge expressions," themselves internal, between these mechanisms and systems. According to the terminology proposed by Hauser et al. (2002), we study language in a "broad sense" when we adopt the communicative point of view; whereas we study language in a "narrow sense" when we adopt the computational point of view. In this narrow sense, then, language is *not* a communication system; but it *is* one in the broader sense. In any event, the computational system—or Faculty of Language in the Narrow Sense (FLN)—is one more piece of language as a communication system or Faculty of Language in the Broad Sense (FLB).

One of the main claims of Hauser et al. (2002) is that the computational system used by FLB appears to be unique, with no equivalent in other aspects of animal cognition (including human cognition, with language itself being the exception). The kinds of operations FLN is capable of executing have a degree of complexity well over that of "Type 2 grammars" in the Chomsky Hierarchy—which makes

---

[8] On the ("minimal") capabilities of dealing with human syntax by non-human primates in experimental contexts, see Premack (1985) for the case of chimpanzees, and Savage-Rumbaugh et al. (1993) and Savage-Rumbaugh et al. (1998) for the bonobo. Experiments such as those reported in these works are taken by primatologists to be the strongest evidence for evolutionary continuity between the communicative capacities of primates and language.

them apt therefore to establish properly recursive hierarchical relations, while nothing in the study of the communicative behavior of other animals suggests levels of complexity beyond "Type 3 grammars"—finite-state systems only apt to establish strictly linear relations. As a consequence of that, the computational system would be a genuinely new incorporation to FLB and would hence count as the "key innovation" in the context of the evolution of the communicative function (Hauser et al. 2002: 1573).[9] The very same level of computational complexity might however be present in the characteristic operations of other, non-communicative aspects of animal cognition, such as spatial navigation, number quantification, or social relationships (Hauser et al. 2002: 1578), but this should not obscure the fact that FLB is a highly innovative system of communication, since even though the computational procedures it incorporates would not be exceptional in themselves, they would in any case be entirely alien to any other form of non-human communication. Hauser and his coauthors also point out (2002: 1573) that, from the point of view of the evolution of communication, another important novelty is represented by the "interfaces" or contact areas between FLN and the other systems of FLB, which give rise to a high degree of generality to the kinds of externalizable contents (as opposed to the specificity and domain-centered nature of other animal communication systems), and to a relative flexibility as to the available externalization mechanisms (vocal-aural, visual-manual, again in contrast with non-human animal communication).

Hauser et al. (2002) is often read and interpreted as a defense, for the reasons just noted, of the specifically human and specifically linguistic character of FLN, the computational aspect of language. This interpretation, we believe, is correct; but only partly correct. This work also, and perhaps above all, is a defense of the singularity of language qua communication system and, in this sense, a particular application of the research program on the evolution of the communicative function developed by Hauser (1997), whose explicitly declared main goal is to determine the causes underlying variation in natural communication systems, human linguistic communication included. Indeed, one should read as a direct appeal to this framework the fact that one of the basic assumptions of Hauser et al. is that, despite the conspicuous discontinuity among the systems on which communication is based in different species, these systems are nothing but several instantiations of a unique but highly diversified organic function (2002: 1569; especially, Fig. 1). The paper constitutes, then, an attempt to provide an answer for the particular case of linguistic communication within a broader research agenda seeking the causal factors responsible for the piecemeal diversification of the communicative function in its evolution within different species (Hauser 1997: 1–2). In this sense, we believe it is also licit to read it

---

[9] On the notion of "key innovation" in the evolutionary sense, see for example Hall (1999: 216–217).

as offering the crux of the matter concerning what sets language—as Chomsky has argued since at least the 1960s—entirely apart from the rest of animal communication. For this reason, and because we are led to interpret that Chomsky himself approves of the Hauser et al. (2002) way of settling the issue, it may be pertinent to remind the reader that Chomsky's main objections for categorizing language along with the other animal communication systems have mostly to do with its essentially unspecific nature from the functional point of view: With the different manifestations of animal communication being "purely functional" and "stimulus oriented" (Chomsky 1966: 30), language is characterized instead by the absence of any kind of "functional specialization" (Chomsky 1975: 111, 1980: 240; also Lorenzo 2008: §2.2, for an overview of Chomsky's anti-functionalism). Hauser et al.'s (2002) proposals concerning the singularity of human language in the broader context of animal communication articulate an answer to the previous objections along the following lines:

1. The diversification of the interfaces between FLN and the other systems of human thought (or, alternatively, the diversification of the interfaces among these systems and the establishment of an interface with FLN) gave rise to the multifarious functions subserved by language; and
2. The access of FLN to a properly recursive computational regime (i.e., with an unlimited capacity for embedding chains of symbols within chains of symbols) endowed language with a similarly unlimited expressive power, disproportionate for the satisfaction of any homeostatic need.

Thus Hauser and his coauthors apparently manage to "normalize" the "exceptionality" of language with respect to the evolution of the communicative function in strict compliance with the research agenda of Hauser (1997).

3.2.2 Language is not an animal communication system...
and neither are animal communication systems

There is an aspect of Chomsky's anti-functionalism which none of the approaches reviewed above manage to answer in any obviously satisfactory way, precisely because both Chomsky's and the other approaches are constructed on the same fallacy. But this is the core element of Chomskyan anti-functionalism: The idea that language is not good for any specific function *in particular*, beginning with the communicative function itself. Taking into account that strictly individual uses are as consubstantial to language as collective uses, that expressions may be externalized or not with the same degree of naturalness, and that externalized expressions may or may not have a truly informative value, without any of these circumstances making them more or less linguistic, Chomsky concludes that language cannot possibly fall within any definition of "communication system" we may think of (Chomsky 1968:

123).[10] All things considered, accepting that language is an element of the so-called "communication systems," as is generally assumed both in the Darwinian and the Hauserian versions of the communicative fallacy, looks more like a stipulation inspired by the availability of the notion "animal communication" than a rational decision based on evidence that language is a true member in the class of animal communication systems.

As a matter of fact, this situation is, in our opinion, much more serious and deserving of a more severe criticism than that expressed by Chomsky. It is our contention that language is not, and cannot be, an animal communication system because no animal communication system is in fact an animal communication system. The reason is clear: The concept "animal communication" does not designate a natural kind.[11] As a consequence of that, any variety of the communicative model suffers from a very serious, foundational problem. In the end, the diversity and discontinuity in the so-called animal communication systems is easily explained by the fact that no natural unity exists among these systems. Therefore, the diversity that, for example, Hauser (1997) tries to explain is not the result of some evolutionary radiation of varieties of the same natural phenomenon whose causes call for an

---

[10] With the formalization of the notion of "communication" carried out by the mathematical theory of information (Wiener 1948; Shannon and Weaver 1949), this observation becomes even more appropriate, but in the opposite sense. This theory focuses on the ability of signals to reduce uncertainty in specific situations and on the importance of such mechanisms as redundancy or feedback in guaranteeing the successful transmission of the information contained in the signal, which appears to be ancillary to the actual interchange of signals between non-human animal agents. It must also be pointed out that Shannon and Weaver were particularly cautious to dissociate their mathematical models from any theory of meaning or interpretation, aiming at confining them strictly to electrical engineering: "...semantic aspects of communication are irrelevant to the engineering problem" (Shannon and Weaver 1949: 3). According to the critical examination of information theory carried out by Bar-Hillel (1955), Norbert Wiener is to be held responsible for the confusion between "quantity of information" and "quantity of meaning."

[11] For a number of reasons, some readers might argue at this point that our use of the term "natural kind" is unfortunate here if not potentially damaging for the arguments to be developed in the text. One of these reasons, some might say, is that we are attributing an essentialist thinking to a framework that is not essentialist at all, because, as Ernst Mayr made clear long ago (e.g., in Mayr 1963; see Sober 1980, for an overview), Darwinism replaced typological thinking with population thinking, and we are therefore attacking a straw man. Another of these reasons, some others might say, is that we are projecting our own essentialist thinking onto a framework that is not essentialist and form here the very same criticisms as would have been applicable before. Our answer would be, in any of these cases, that essentialism, as it is traditionally understood, is not at issue here (and thus, we might concede, ours is perhaps not the best of all terminological choices), but rather the idea that *there exist objective criteria for the individuation of functions*. And this *is* certainly an integral part of the essence of adaptationist thinking; that is for adaptationism to work at all it is necessary that for every phenotype (behavioral, morphological, whatever) P a (proper, main, primary) function F exists for which P is designed and, crucially, F can be individuated on objective grounds. In other words, what we require is an objective criterion for saying that, for example, behaviors $b_1, b_2, \ldots, b_n$ have the same function F (e.g., communication), independently of their form; similarly for morphologies. This idea that behaviors, morphologies, whatever can be grouped together only on the basis that they have the same function strikes us as an essentialism of sorts, a functional essentialism that assumes the existence of natural kinds of functions. We'll come back to this issue from a slightly different perspective in Chapter 4.

explanation; and the argument supporting this standpoint is not difficult to construct, while it may be hard to accept, given the fact that the concept of "communication" is so popular. Let's take it one step at a time.

First, it is important to understand that the concept "animal communication" takes us back to the kind of functionalist explanation typical of ethological models, whose logic, at least in this preliminary phase, is going to drive our own criticism. Let us then accept, for the sake of argument, the logic of the functionalist explanation; we will shed it in the following section. Modern ethological tradition collects under the tag "animal communication" all forms of behavior implying some *contact* between organisms through *signals* made *public* by some or other means, and to which some *informative content* and *usefulness* can be attributed on the basis of its putative connection to the homeostatic equilibrium of the organisms in question. The defining character of the italicized terms in this formulation of the concept is explicitly stated, for example, by Smith (1977: 11–34), a classical perspective on the topic, which we may safely take as representative of the aforementioned application of the ethological tradition to the particular case of communicative behavior. It is also interesting to emphasize the fact that such terms match those on which Chomsky based his thesis of the inappropriateness of linking the scientific study of language to the analysis of the so-called "animal communication systems." As already pointed out, these terms cannot really be seen as defining, in the case of language.

As far as our argument is concerned, we want to focus on the last of these putatively defining terms, according to which every system of communication corresponds to some characteristic kind of *usefulness* associated with the environmental and populational economy of the organisms employing the system in question. It is a particularly important feature for the ethological model, because, as pointed out by Smith (1977: 16), it has the power to confer an adaptive value to communicative behaviors, and it justifies our seeing them as products of natural selection. Should we wish to resort to more modern terminology, we would say that communicative behaviors and the signal systems on which they are based are held to possess a *causal function*, that is, a relevant role in the homeostatic regime of its practitioners, which eventually refers to a *selective function*, that is, a historically favorable balance in the reproductive rates of these organisms relative to organisms not instantiating these behaviors and these systems (Godfrey-Smith 1994; Griffiths 1994).

Let us not yet discard this functionalist logic. Several species of Old World monkeys call in order to alert their conspecifics of the presence of some predator (Struhsaker 1967; Cheney and Seyfarth 1990: Ch. 4). It goes without saying that being alerted is *useful*, whereas not being alerted when some peril is imminent is not; nor can one question the reproductive benefit derived from the possibility of extending one's life a bit more. Males of most avian species sing to attract the attention of females to their qualities (Collins 2004); so one is wrong to deny the benefits of being able to seduce a female, or the reproductive advantages derived from being the winner in such seduction games. We might easily multiply such instances, but it is preferable to stop here, since the examples are too numerous and the ones presented seem to suffice. Calling and singing play, for monkeys and birds, a fundamental role in the homeostatic equilibrium in each family of species and, clearly, translate positively in the reproductive rates historically attained by the most successful practitioners of such behaviors. Once again, we come to the point that we think calls into question the ethological definition of animal communication: If the calls of *Cercopithecus aethiops* and the songs of *Fringillia coelebs* are, causally and selectively, natural alert and seduction systems, what else adds their further attribution to the communicative function?

Our question is rhetorical and the answer is clearly nothing. Everything that can be said about these systems from the causal and selective point of view has been exhausted by their characterization in terms of natural alarm and seduction systems. Nothing else needs to be added for these systems to be exhaustively characterized from the functional point of view. Adding the communicative function on top of all that is a mere redundancy, and in the end renders a functional characterization of animal behavior to be entirely useless in all cases.

The case of "communication" is not unique. It is just a particular case of many other ordinary language terms that cannot survive their transfer to the vocabulary of a theoretical discipline (biology in our case), because they do not denote anything made up of a natural kind. For example, Paul Griffiths carried out a detailed analysis of this phenomenon with the concept "emotion" in psychology (Griffiths 1997, 2004a, 2004b), and his conclusions can be imported, word for word, to the case of "animal communication" in biology: In both cases, the range of application of the terms is so wide and heterogeneous, and the concurring properties in each case so diverse, that one cannot expect any reliable extrapolations from putative token instances of the category to extend to the category as a whole. This means that such terms do not implement natural kinds to which either logic or the practice of scientific discovery can be applicable. The case of "animal communication" is particularly representative in this since, as well as the aforementioned alarm and seduction systems, systems for dissuasion, spatial representation, deception, etc., should also be added which, as Griffiths claims with the case of "emotions," would be better studied independently rather than as instantiations of some superordinate class.

In other words, when we speak of "animal communication" we are not referring to a natural kind, but to a highly heterogeneous collection of behaviors for which there is no hope of finding a sufficient and satisfactory set of common general principles in order to justify a research program. Their reference is "partial:" It refers "in part" to alert systems, "in part" to seduction systems, "in part" to deception systems... just as partial is the reference of many other ordinary language terms, which are useful for a pre-theoretical approach to some theoretical domains, but whose reference cannot ever attain a subjacent natural identity. The notion of "animal communication" appears to deserve the same kind of eliminativist pruning that, for some philosophers, is deserved by other ordinary language terms, like, for example, "consciousness," routinely applied in psychological theory (Churchland 1981, 1985; Dennett 1991, among others).

Nothing in what is usually called "animal communication systems" is, biologically speaking, an animal communication system. However, to help get a clearer picture of how what we have said affects the theoretical underpinnings of the evolutionary study of language, we continue our criticism, with the final goal of achieving a definitive undermining of its current functionalist foundations.

3.3 Against the Functionalist Fallacy

None of the points we have touched on so far constitutes, however, the main problem for the underpinnings of those frameworks that are based on the communicative fallacy. Intensifying the critical tone of our analysis, we shall point out that a much more severe problem than those discussed so far is the use of the notion "communication" to refer to a phenomenon actually independent from the organic structures in which it is manifested and toward which, given its indisputable advantages, the evolution of any complex organic form seems somehow inevitably directed. It is a conception of communication as an evolutionary a priori, giving rise to a position where function not only acquires explanatory priority over form but also becomes an independent, transcendental phenomenon whose existence does not presuppose the existence of any form in particular. As this is a problem affecting not only those models seeking an evolutionary explanation of language using the concept of "communication"—models instantiating a variety of the problem we could qualify as "communicative transcendentalism"—but any evolutionary model whose mode of explanation is strongly based on the notion of "function," we may broaden the range of our criticism in order to include such models. We will refer to this mainstream stance in contemporary evolutionary thought as "transcendental functionalism," which we think constitutes an insurmountable problem for neo-Darwinian adaptationist approaches.

Think, for example, of the particular case of attributing the function "alarm" to the calls of the different species of Cercopithecidae. It is interesting that, with respect to this particular case, a number of species with no close evolutionary connection to these monkeys, such as some species of birds like the Siberian jay (Griesser 2008), have developed similar systems of vocalizations discriminating different types of predators and playing the role of collective alarm signals. Now, the act of attributing the *same* function to these behaviors necessarily implies granting this recurring function an existentially autonomous character and the possibility that it be instantiated in many different species and organic systems; similarly with seduction, spatial representation, or any other function attributable to some collection of organic structures, including the queen of all functions (or, at least, of all these ones), the communicative function. This is so because the same criticisms concerning the impossibility of establishing a natural kind capable of including all "communicative functions" are directly applicable to, say, the functions "alarm" or "seduction." Let's take the second of these functions, "seduction," as an example: we immediately see that what typically falls under the label of "seduction" or of "courtship" is in fact a heterogeneous collection of behaviors such as courtship feeding (Nisbet 1973), territoriality (O'Donald, 1963; Wynne-Edwards 1962), pheromone emission (Thornhill 1992; Moore 1988), construction of artifacts or shelters (McKaye 1979; Christy 1988; Borgia 1986), display of some body part (Petrie et al. 1991), or performance of some kind of dance (Gibson et al. 1991), to which we could add vocalizations, of course, plus some other examples, such as the possession of a brightly colored pigmentation or a very long tail (Møller 1994), none of which could really be classified as behaviors and which Hauser (1997) includes within the class of "cues." The point is that almost none of these behaviors or characters are exclusive to "seduction," such as ritual dances, which can be associated with "fight" behavior; see Zahavi and Zahavi (1997) for an overview of these and other examples.

All this boils down to the fact that the metaphysics of functions is sufficient to deprive them of any authority when individuating organic structures and of establishing relevant identities ("homologization," see Chapter 4) to construct evolutionary explanations. As we have already noted in connection with the communicative function, no function names a true natural kind (they name, perhaps, something like transcendental functions), and, as Fodor and Piattelli-Palmarini (2010: Part II, Ch. 3) point out, no contingent explanatory law exists if it is not based on actual natural kinds. Besides that, "transcendental functionalism" resorts to a methodology based on a top-down strategy, taking, as we've just argued, a set of "transcendental" and unmistakably "anthropomorphic" categories that are later imposed onto behaviors that, to our eyes, appear to fit them perfectly. Thence the practice of talking about "infanticide lions" (Pusey and Packer 1994), which kill the cubs they did not father, or "homosexual male penguins" (Ebeling and Spanier 2011), which take care of an abandoned chick. A few examples like these suffice to show that this form of functionalism is also fallacious, because it rests on categories which are the product of a human mentality and whose transcendental or impositional character derives only from a tendency, also typical of human mentality, of anthropomorphizing nature.

But functionalism may be dismissed on other grounds, and the case of the evolutionary characterization and explanation of language in functional terms is a particularly good point of departure to criticize another aspect of the "functionalist fallacy" which, we believe, is very important to rebut. Let's return now to the Chomskyan argument that no use of language exists that may be considered as characteristic: We use language publicly and privately, to convey information or to deceive our fellows, to reveal or clarify states of uncertainty, with no full guarantee, or actual interest whatsoever, of being understood... Language, as often emphasized by Chomsky (1975: 111, 1980: 240), lacks a "special function," and this is perhaps the only convenient characterization we can come to by making use of a functional vocabulary. It is not good for anything in particular and is in fact fairly useful for whatever we may want to do with it.

Different authors have at some point tried to pinpoint one or another of these functions and, on the grounds of diverse criteria, have identified the function chosen as the special function of language. In most of these cases, this function is put into some relation to or directly identified with the so-called "social uses" of language, perhaps as a sign of the attachment these authors feel to the communicative fallacy we dismantled in the previous section. For Aitchinson (1996), such uses are those in which we show a higher degree of verbal fluidity, whereas Dunbar (1996) suggests they are those we spend most time with in our conversations. According to this line of argumentation, any other uses we make of language, even if consubstantial to its effectiveness, are at most just "parasitic uses" in lieu of this essential function.

The problem with these approaches is (at least) twofold. First, criteria such as naturalness, frequency of use, or any other putatively useful criterion to determine what the special function of language is, are established on strictly stipulative grounds: It is not clear what the valid criterion could be to select the valid criterion (and so on, in an infinite regress). Last, but not least, the de facto association of language with a manifold of possible uses or functions poses a problem that Fodor and Piattelli-Palmarini (2010: Part II, Ch. 2) identify as endemic to any kind of explanation appealing to the different causal roles of co-extensive properties. To pinpoint as special a single function from a set of functions while relegating the rest to the status of parasitic requires the use of counterfactuals, that is, of possible alternative situations capable of showing the preservation of the privileged function in the absence of any of the others. But language, let us insist once more on this point, is de facto associated with a manifold of possible uses and functions, and "de facto" here means that language is an organic system that evolved in the world of contingency, within a very specific, and perhaps unique, context, which automatically makes the mechanism that favored its emergence entirely insensitive to counterfactuals. As Gould (1989) so effectively expressed it, if we were given the chance to rewind the tape of life and replay it again, the images and sounds of evolution would adopt completely different forms and hues from the ones it actually adopted. There is no point, then, in establishing any type of distinction between what might be considered as essential and what might be seen as accidental in life in general and in language in particular. The different uses we can make of language are all equally essential or, for that matter, accidental, and consequently language is characterized by an essential lack of functional specificity.

Moreover, it may well be the case that what appears to be true of language turns out not to be as exceptional as we might think. The call of a vervet monkey, for example, serves also to signal its position within the hierarchy of the group (Cheney and Seyfarth 1990); male finches also sing when they are alone, and, at those times, whatever the purpose of their singing, it is certainly not to seduce a female (Teramitsu and White 2006). Many similar examples could be given. Because if what deserved the unfortunate name of "animal communication systems" appears to possess such a high functional specificity, it may be because of our own tendency to see them like this, as a reflex, no doubt, of a more general tendency to see in the morphological and behavioral traits of animals examples of their "exquisite adaptation" to the environmental and populational conditions they live in. Should we free our observations of animal behavior from prejudice, we would probably see "parasitic functions" proliferate and eventually end up questioning the qualification of "essential function" regarding any of their manifestations. Concomitantly, more judicious interpretations of the causal role of any of the traits making up a phenotype (the large bill of the toucan *Ramphastos toco*, for example, to cite a case that deserved some attention recently, and which we discussed in the previous chapter in a slightly different context; see Tattersall et al. 2009) would not be resolved by eliminating or despising one function (sexual attraction) once a new and, putatively at least, equally important one is discovered (thermoregulation). It appears to be equally as likely that the trait in question serves one or the other function as that it in fact serves both (or more) at the same time, in which case, and independently of their relative importance for the ecological economy of the species (even assuming that this can be effectively determined), they will be de facto mutually parasitic and evolutionarily inseparable functions.

The conclusion that language lacks a special function should not come as a surprise then. As argued in Balari and Lorenzo (2010a) the conclusion merely maps onto language something that is actually valid for some different kind of organic structure: All organic structures are functionally non-specific, beyond our natural proclivity to see them as artifacts designed with some specific purpose (Searle 1995: 41; also Ruse 2003, although drawing radically different conclusions).[12] Indeed, organic structures possess certain formal or structural properties which, given the organic

---

[12] Functions are, therefore, in the mind of the beholder and, hence, there are no *objective* criteria for functional individuation, only subjective ones. It is interesting to note in this connection that, at least since Larry Wright's (1973) definition of the selected-effect concept of function, the only criterion for the individuation of functions to which philosophers of biology have appealed is natural selection itself. This position has led some authors to defend some quite outrageous ideas in order to "save" functions and adaptationism. Thus, for example, Dennett (1987: Ch. 8) is willing to accept that, of course, functions (and meanings) are in the mind of beholders but that natural selection is one of them, only that it is a beholder without a mind. We confess ourselves totally unable to understand this. We cannot enter into a detailed analysis of the problems posed by the selected-effect concept of function, but see Balari and Lorenzo (2010a).

and environmental context in which they integrate themselves, endow them with a more or less varied practical potential and, also, a quite variable one from one case to another. Balari and Lorenzo (2010a) call this property of organic structures "functionability," emphasizing the fact that it is a strictly formal notion. Let us dwell a little longer on this notion, as it appears to play a crucial role in an evolutionary explanation of an organic system like language.

Balari and Lorenzo (2010a: 68–70) define the "functionability" of a structure as a generic functionality of sorts, that is, as something like the operative elbow room left to that structure by its formal properties and which can give rise to a more or less varied gamut of uses and practical benefits. The authors also contend, however, that organic structures are not functionally specialized per se, rather that they simply possess no characteristic function or set of functions in particular. Obviously, this is not to say that they are useless; it just means that their usefulness derives from the formal properties of their structure, from the organic context of which they are an integral part, and from the environmental circumstances in which the organism manifesting them is localized. Organic structures are not characterized, in sum, by their *(practical) functionality*, but by their *(formal) functionability*. Those cases of structures with a seemingly high or extreme specialization for a single function or task just constitute one of the extreme cases of functionability, where their formal properties leave a very narrow margin of operability. At the other end of the spectrum, some organic systems possess a set of formal properties that make us see them as entirely free from any practical specialization. In any case, what characterizes these structures is not their apparent function or lack thereof, but their *aptitude to function* according to the formal features they possess.

All this bears upon the need, emphasized by Love (2007), of distinguishing between the proper *activity* of an organic structure (its functioning), determined by the internal organizational properties of the structure in question, and the actual *use* to which it can be put given the organic and environmental context in which it is realized (its usefulness); see Wouters (1999, 2003) for some finer distinctions in the characterization and individuation of different types of functions. Thus, as Love explains, while the notion of "activity" refers to (or is identifiable with) "*how* the structure is," the notion of "use" refers instead to "what the structure is *for*" (Love 2007: 695–696). In this context, the notion of "functionability" may then be defined as the degree of flexibility in the activity of an organic structure, and this is a strictly formal (or, following Love 2007: 701–702, "structural") property and totally underdetermined with respect to the use or usefulness the said structure may offer to the organism that developed it.

Finally, Balari and Lorenzo (2010a) reason that the concept has a clear projection onto the evolutionary arena, in the sense that a high degree of functionability—a high capacity to respond in unexpected ways to organic change and environmental variation, may be one of the factors guaranteeing the evolutionary success of a given organic structure. This suggests a connection of the notion of "functionability", which refers to the actual organizational state of an organism, with the notion of "adaptability" as defined by Reid (2007) in contraposition to the classical notion of "adaptation." As Reid (2007: 13, 241) explicitly notes, a high degree of adaptation to specific environmental conditions may turn out to be an evolutionary catch-22 given the essential instability of the medium; on the other hand, a high capacity to formally reconfigure oneself may open up for an organism a whole new set of avenues to respond to environmental changes or move to new environments. Balari and Lorenzo (2010a: 69) conclude, then, that *"functionability, that is, not being adapted to any environment (medium or context) in particular and having the ability to function in new and unexpected manners is one of the keys to evolutionary success."*

Thus, in a broader context, functionability may be seen as a species of plasticity, according to the sense we defined it in the previous chapter, but with a special emphasis on the activity side of the matter and, hence, specially suited to dealing with change at the physiological/neuroendocrinological (and behavioral) level, which may in turn be correlated to morphological change. This suggests that plasticity may not be a unitary phenomenon, that it is one thing at the activity level (i.e., functionability) while another thing at the morphological level; both are mutually and causally related in the sense that morphological plasticity and change may alter the degree of functionability and so foster (or restrain) further change at the activity level and vice versa. This idea will be of some importance later in this book, when we discuss brain evolution and development.

3.4 Conclusions

We are now in a position to present two fundamental cues to guide the evolutionary study of language, whose success may be put in direct correlation to its high functionability and the corresponding degree of adaptability it bestowed on the organisms in which it evolved. First, it makes no sense to base an evolutionary explanation of language on the attribution to such an organic system of any kind of functionality in particular or on the association with any kind of adaptive advantage for the speakers. Second, an evolutionary explanation of language requires establishing what formal readjustments from some common developmental ground may have taken place in order to give rise to an organized structure with a high degree of functionability. We clarify this latter point, by stressing that although animals may do something in many different ways, one way turning out to be more or less useful than another or others, we deny that such usefulness is causally, epistemologically, and evolutionarily prior to the structure that made it possible in the first place. Behavior is not plastic, only the structures that cause it may be. Novel manifest forms of behavior can only arise if the appropriate conditions of adaptability and functionability have previously been attained. Such conditions are eminently structural (morphological, neurological, physiological, and, as we will argue in Chapter 5, computational), and we must turn our attention to these if we want to apply the comparative method in order to understand the origins of biological (including behavioral) novelty. An important corollary of these propositions is that everything that is observable in the behavior of animals is transparent with respect to the underlying biological structures subserving it and that a formal analysis of this behavior, independently of functional considerations, may shed light on the formal properties of the said biological structures, thus enabling the application of the comparative method at these structural levels. Language is no exception to this principle, and this insight will drive the articulation of our proposals from Chapter 5 onward, but we need first to devote some attention to "homology," the central notion for the application of the comparative method in biology, a task we undertake in the following chapter.

4

On True Homologues

Et si les parties qui diffèrent le plus en apparence se ressembloient au fond, ne pourrait-on pas en conclure avec plus de certitude qu'il n'y a qu'un ensemble, qu'une forme essentielle, et que l'on reconnoît partout cette fécondité de la nature qui semble avoir imprimé à tous les êtres deux caractères nullement contradictoires, celui de la constance dans le type et de la variété dans les modifications?[1]

Félix Vicq d'Azyr, 1774

Naturgeschichte beruht überhaupt auf Vergleichung. Äußere Kennzeichen sind bedeutend, aber nicht hinreichend, um organische Körper gehörig zu sondern und wieder zusammenzustellen.[2]

Johann Wolfgang von Goethe, 1795

In the previous chapter we developed an argument for the primacy of form over function that was mostly based on the demonstration that functions cannot be considered to be natural kinds as there do not appear to exist any objective criteria to ground them. Thus, we saw that functions cannot be grounded on structure because in the vast majority of cases structures are not transparent with respect to the kinds of functions they can perform (i.e., from the inspection of a structure one is hardly ever able to infer the function it subserves) and, moreover, structures tend to be functionally plastic or multifunctional (i.e., they show high levels of function-ability). As we saw, the only remaining alternatives to preserve the integrity of functional classes are either to ground them on natural selection or resort to what we called "transcendental functionalism," which, we argued, are ontologically

---

[1] "And if it were the case that those parts that appear to be more different from each other were in fact fundamentally alike, couldn't we then conclude with greater certainty that there exists just one unity, an essential form, and that everywhere can we recognize this fertility of nature, which appears to have imprinted on all beings two characters that are by no means contradictory, that of the constancy of type and that of the variation under modification?" (Our translation, SB and GL.)

[2] "Natural history is mostly founded on comparison. External features are important, but not sufficient, to accurately break up organic bodies and to put them back together again." (Our translation, SB and GL.)

untenable for several reasons. An immediate conclusion of these criticisms was that the comparative method in biology can only be applied to structures, for, in this case, there are criteria to define classes within which comparisons can be established independently of functional considerations.

In this chapter, we pick up this conclusion in order to complete our defense of the primacy of form over function by focusing our attention on homology, one of the central notions underlying all comparative work in biology and which, ever since Darwin, has often been detached from its original structural/formal basis in order to extend it to the domain of function and behavior.

At the onset of his review of Hall (1994), David Wake wrote that "Homology is the central concept of *all* biology" (Wake 1994: 268; emphasis in the original), and then he went on to conclude that "It is sufficient to 'know' that homology, like truth, exists." These two quotes by Wake epitomize what may well be taken as the most common attitude adopted by biologists towards homology: It is basic, we don't have a clear definition of it, but it works. Indeed, one can hardly deny the latter and, since its origin in the field of comparative anatomy, homology has been fruitfully applied at several other levels of analysis such as the genetic, molecular, and developmental ones, to name just a few; see the papers in Hall (1994) and in Bock and Cardew (1999) for an overview. There is probably one exception to the rule, however, and this exception is behavior. In fact, the applicability of some criteria of homology to behavior remains controversial and most of the literature devoted to the topic still circles around Atz's (1970) original assessment against importing homology to the realm of behavior. Leaving aside for the time being the validity of Atz's arguments, the truth is that the idea of behavioral homology has proven so far much less fruitful than its counterparts at the genetic, molecular, or morphological levels and, when it has been applied, it has been to relatively stereotyped (we could even say "hardwired" in many cases) behaviors.[3] Therefore, complex behaviors or, as we would rather call them for reasons to be made clear presently, higher cognitive capacities have so far resisted an analysis in terms of behavioral homology.

In this chapter we would like to take up this issue by reassessing the status of homology at the level of cognition. This is a necessary first step before proposing in the next chapter a criterion of homology that, in our opinion, may prove useful for extending the comparative method to this level and thus fill this gap in the hierarchy of comparative biology. For the time being, we will expound a number of arguments devised to explain the relative lack of success of the notion of "behavioral homology." Some of these arguments concern mostly the "behavioral" part of the term, but some others concern "homology" in general and are therefore relevant for its application in

---

[3] See Wenzel (1992, 1993) and Greene (1994, 1999) for some representative examples; also Chapter 8 for a more precise characterization of what we call here "stereotyped behaviors" and "higher cognitive capacities."

biology as a whole. As for the former, our contention will be that the problem with the idea of "behavioral homology" is its focus on behaviors as the only relevant data in the study of cognition, a methodology whose principles are directly connected with behaviorist psychology, but which is untenable in contemporary (computational) cognitive science. As for homology per se, on the one hand we can only agree with Atz (1970) when he wrote, commenting on Simpson's (1961: 78) definition of homology as "resemblance due to inheritance from a common ancestor," that "Simpson's definition implies another extension of the original meaning of 'homology;' it applies to functions as well as structures" (Atz 1970: 53), but, on the other hand, we will also agree, for example, with Boyden (1969) that this constitutes a perversion of the meaning of homology as understood by Owen (1848) and that Simpson's (and, as a matter of fact, Darwin's) redefinition of homology cannot be sustained in part because of its consequences for the analysis of behavior already pointed out by Atz. Our conclusion will be that something like the notions of Owen's "special" and "general" homology are *both* necessary for the applicability of the comparative method to biology in general and to the biology of cognition in particular.

4.1 On Behavioral Homology

The very term "behavioral homology" already reveals a strong bias toward the idea that to apply the comparative method it is sufficient to pay attention to what animals do (preying, burrowing, communicating, and so on). This idea is based on a number of assumptions that, as we will try to show in the following paragraphs, suffer from a number of methodological shortcomings that seriously undermine its actual applicability. As a consequence of that, we will contend in the next chapter, the study of animal behavior in general and the extension of homology to this realm needs to undergo a paradigm shift comparable to the one carried out in the cognitive sciences during the 1960s with the so-called "computational revolution."

Following our line of criticism of behavioral homology we will eventually come to call into question, on independent grounds, the very notion of historical homology that made behavioral homology possible. We address this issue in the next section, where we present a number of arguments for restoring the original, ahistorical or, as we would like to name it, Owenian definition of homology.

The idea of homologizing behaviors is based on the following, more or less explicit assumptions, shared with more or less radical varieties of behaviorism:

1. Observable behaviors can be classified into natural kinds.[4]

---

[4] Strictly speaking, this assumption is not shared with behaviorist psychology, which acknowledged the fuzziness of behavior but nevertheless maintained the possibility of distinguishing between *central* and *peripheral* instances within a specific behavioral class. We come back to this issue below.

2. Action can be explained by appealing to a disposition to act in a specific way and, consequently, that knowing something can be reduced to knowing how to do something.
3. A general theory of learning is sufficient to explain how an organism comes to know how to do something.

In what follows we will examine these three assumptions with the aim of showing that all three are entirely unwarranted. Most of our arguments are familiar from the literature of contemporary cognitive science, but we believe it important to rehearse them here to make a stronger case for the alternative we will be presenting below.

4.1.1 The fuzziness of behavioral categories

These three assumptions are closely intertwined, but we'll try to discuss them in turn. Thus, and starting with assumption 1, any claim for the homology of any two entities needs to be based on some previously established frame of comparison. This is relatively unproblematic in the case of structures, since we can appeal to such criteria of similarity as form, shape, connections, etc., familiar from earlier work in comparative anatomy by Vicq d'Azyr, Geoffroy, Goethe, Owen, and many others. These criteria, however, do not have a straightforward translation to the realm of behavior, since, crucially, in this case some notion of function appears to play a significant role in the equation and functional considerations easily yield to paradoxes in the comparative analysis of behavior. Thus, two behaviors following a similar or identical action pattern (e.g., scratching one's ear) need not have the same function, and behaviors with putatively similar or identical functions (e.g., attracting a possible sexual partner) need not necessarily follow the same action pattern. This, in our opinion, already seriously undermines any attempt to construct a valid framework for the application of homology to behavior, since most attempts so far at applying the comparative method at that level follow one or other criterion or a combination of both.

Take, for example, the case of "construction behavior," which has been studied in great detail by Mike Hansell (Hansell 2005).[5] There certainly is a perspective from which construction behavior may be seen as a kind in the sense that all instances of this behavior result in some sort of structure: A nest, a burrow, a trap, or a tool. Now, note that these structures may be grouped into different classes if we perform a finer

---

[5] A caveat. The following discussion is not to be taken as a criticism of Hansell's exceptional study of animal architecture. On the contrary, Hansell is extremely careful at the time of extrapolating his findings to phylogeny and evolution and his work is better understood within the framework of ecosystems ecology (Jones et al. 1994, 1997). In fact, in the opening of his chapter on evolution Hansell states (2005: 226) that "[t]he evolution of animal building can be studied by fitting the respective structures onto a phylogeny established on other criteria." This claim, which will be important later in our discussion, already suggests that there is some circularity underlying the notion of "behavioral homology," since reliable comparisons at this level can be done *once phylogenetic relations have been established*.

grained functional analysis. Thus, both the burrow and the nest may be characterized as shelters, whereas the trap and the tool may perhaps be characterized as implements to catch a prey. There are many possible ways to carve the functional space of construction behavior. As an alternative, we could then concentrate on the actions carried out to build some structure and talk about piling up, weaving, folding, spinning, etc., but, as it is clear in the chapter Hansell devotes to this topic, no interesting phylogenetic or evolutionary generalizations may be derived from this analysis either. In this connection, it is also relevant to note that no anatomical generalizations may be derived from the fact that different organisms perform similar actions during their respective construction behaviors. Thus, Hansell (2005: 93, 95) concludes:

[T]he parts of the anatomy used in construction are overwhelmingly legs and mouthparts (mandibles, beaks, and jaws), supporting prediction 1 of Section 3.1, that construction anatomy would generally be used in other contexts. [...].

Legs and mouthparts may be more or less specialised for their non-building roles, however, this chapter reveals a general lack of specialised anatomy for building.

This conclusion reminds us of the words of Richard Owen who, in a similar context, wrote (Owen 1849: 10):

Nor should we anticipate, if animated in our researches by the quest of final causes in the belief that they were the sole governing principle of organization, a much greater amount of conformity in the construction of the natural instruments by means of which those different elements are traversed by different animals. The teleologist would rather expect to find the same direct purposive adaptation of the limb to its office as in the machine.

Owen's (and Hansell's) message is clear: There are no structures designed to perform specific functions, or, to put it differently, there are no behaviorally specialized organs. We are using the terms "function" and "behavior" somewhat interchangeably here, and this might be taken as a sloppy practice, potentially damaging for our analysis. Such sloppiness does not stem from our way of looking at the problem but dwells at the core of the problem we are discussing here.[6] Thus, picking up again our brief foray into the topic of Chapter 3, one might argue that a "function" is the purpose which some structure (or organ) is for, whereas a "behavior" is something that an organism does. In this sense, then, the notion of "behavior" appears to be intimately tied to some structure or another, as it seems to be connected to some idea of "activity" in the sense of Love (2007). "Functions" on the other hand, somehow transcend the notion of structure and can in principle be individuated independently of it. This distinction,

---

[6] See Kennedy (1992: §3.6, §6.1) for references and a general assessment. In essence, Kennedy's point is that there is a pervading confusion among behaviors and the functions of behaviors and that this confusion leaks into the lower levels of analysis. We agree with Kennedy's diagnosis, but as will be clear when we come back to this point presently, not with his cure.

however, is much less clear than it seems. Thus, for example, Hodos (1976) assumes that behavioral homologies can only be established insofar as the compared behaviors depend on homologous structures, whereas Greene (1994) appears to be much less prone to accept such a view but, at the same time, also appears to assume that "behavioral vocabulary" can be reduced to a collection of Modal Action Patterns (MAPs) or "repeatable, spatiotemporal patterns of movements and postures" (Greene 1994: 378). Now, if we go a little deeper into Hodos's and Greene's analyses, we find on the one hand that for the former, given that the wings of birds and the forelimbs of mammals are homologous *as limbs*, "any behaviors that involve use of wings and use of forelimbs are homologous, at least as movements" (Hodos, 1976: 155). On the other hand, Greene assumes that homologies are to be established at the level of MAPs, but only within an independently predefined behavioral category.[7] Thus, Greene (1994), in his case studies, identifies a number of MAPs that he classifies as constricting behaviors and a number that he classifies as gaping threat displays, with both kinds being restricted to a single taxon, namely snakes. Now, note first how in both cases the notions of function and behavior are actually blurred. In Hodos's case we start with a previously established structural homology between avian and mammalian forelimbs and, given that the common ancestor of both birds and mammals also had forelimbs which were for something, we apply to wings and forelimbs the behavioral category which more closely captures what wings and forelimbs are for, and this category happens to be movement. In Greene's case, the behavioral category (constricting or gaping threat display) is defined a priori and only thereafter are the different MAPs identified that are for constricting or for gaping.

We have dwelled on these two cases because, in our opinion, they constitute paradigmatic examples of how the extension of the notion of homology to behavior constitutes a trivialization of the comparative method in biology. Thus Hodos "saves" behavioral homology by associating it with structural homology at the cost of resorting to such "umbrella" behavioral/functional categories as "movement," which adds very little to the comparative analysis established previously on strictly formal grounds. As a matter of fact, Hodos's strategy is an attempt to ground some notion of function on the diverse kinds of structures capable of performing a given function under the (implicit) assumption that *homologous structures perform homologous functions*. Thus, in the case of limbs, the only functional common denominator is the label "movement," educing the conclusion that moving a wing and moving an arm are homologous "as movements." We are in any case left wondering what to do with

---

[7] Greene doesn't say this explicitly; it rather follows from the two case studies he presents in his work, where different MAPs are defined as instances of some specific behavior. This is why we used the word "reduction" in the text, since, and to use one of Greene's examples, a specific MAP observed in some species of snake *is a* constricting behavior, which is independently defined as a "pattern in which prey is immobilized by two or more points on a snake's body" (Greene 1994: 371).

"movement" in snakes, snails, spiders, and fish, since in these cases "movement" is perhaps of a totally different kind from "movement" in birds and mammals, a conclusion that would perhaps have troubled Aristotle, but certainly not Owen.[8]

The case of Greene is different because trivialization comes from a different source, namely the fact that the behavioral category is localized within a specific taxon and subsequent analysis is restricted to that very same taxon. It is not clear, for example, whether Greene would accept as homologous with the MAPs for gaping in snakes the MAPs for gaping behavior in mammals, for example, or even in other reptiles like crocodiles (Kofron 1993). Note that Greene's approach in fact avoids functional talk in the sense of use and comes close to activity talk. The problem is that he does so without actually getting into the real causes of the behavior and limiting himself to the visible patterns of the activity. Greene's approach "works," we contend, only because it focuses on extremely simple and stereotyped activities already circumscribed to a fairly well established phylogeny. This point takes us back to Hansell's words cited in note 5 concerning the reliability of comparisons established at the behavioral level. It is clear that these make sense (if any) only within very restricted domains and within groups whose phylogenetic relations have been established on other grounds and, even in this case, when really complex behaviors are compared the results are most of the time inconclusive, as witnessed, for example, by Hansell's (2000: Ch. 9) own analysis of nest structure in different families of birds. Therefore, behavioral/functional data are perhaps not as useful as it is often assumed at the time of inferring phylogenetic relations (de Queiroz and Wimberger 1993; Rendall and Di Fiore 2007) and, in fact, as pointed out by Sanderson et al. (1993) and by Proctor (1996), these constitute a minimal part of the data used in phylogenetic reconstructions; for example, two recent major works on metazoan phylogeny like Valentine (2004) and Minelli (2009), do not even take them into account.[9] This suggests that

---

[8] Actually, Aristotle would have been more upset than troubled, as he never actually saw the need to ground functions on anything more basic—it is pretty clear that for him functions were primitives. Hodos's failed attempt at grounding functions on structure is a perfect example to show that either one takes functions as primitives (with the ontological difficulties that such an assumption entails) or one tries to ground functions on something different from structure. Most contemporary philosophers of biology are perfectly aware of this fact, concluding either by rejecting the possibility that anything interesting can be said about what some structure is for or by trying to find a viable way to ground functions on history and natural selection, as we have already pointed out in note 12, Chapter 3; see Amundson and Lauder (1994) for an overview.

[9] With one exception in Minelli (2009) corresponding to differences observed in parental care behavior observed in several species of centipedes (Bonato and Minelli 2002). As Alessandro Minelli points out (pers. comm.) it is often useful (but also difficult) to find a strong correlation between some previously established phylogeny and a specific behavioral pattern, and the datum may contribute to strengthen the phylogenetic relationships defined on independent grounds or even to introduce minor adjustments in them on the basis of the observed behavioral data. As pointed out in the text, however, most documented cases of this kind correspond to fairly stereotyped and simple actions and only rarely to more complex behaviors such as nest building; see Wenzel (1993) for an example of this latter case concerning nest building in paper wasps.

the relative success of the comparative method at the behavioral level crucially depends on an indirect connection with previously established structural homologies, which, in the end, results in an approach not very different to the one advocated by Hodos (1976).

Let us summarize our arguments so far. On the one hand, if we adopt the "bottom-up" approach to behavioral homology proposed by Hodos, then the homologies that we establish at the level of behavior, being entirely dependent on previously established structural homologies, are trivial. They are trivial because they hardly add anything new to what we have already established at the structural level: Birds and mammals "move" because their common ancestor "moved." Note that this is independent of what notion of homology we used to establish the structural relation: We could rely either on the historical or the ahistorical one. In fact, the homology of avian and mammalian limbs can in principle be established on both grounds.

On the other hand, if we adopt the "top-down" view of Greene (1994), we focus our attention on a specific taxon (say, snakes) and we identify a behavioral category within this taxon (say, constriction behavior) in order to carry out our comparative analysis. We have already mentioned that Greene's approach is much more damaging for behavioral homology than Hodos's, since the latter merely trivializes the notion, but the former appears to kill it. Let's see how. The crucial point here is that the comparative method is applied within a previously established phylogeny and that the historical notion of homology is explicitly assumed. The second point is crucial and we postpone our discussion to the next section where we present our arguments against the historical notion of homology. As for the first, we believe it is methodologically questionable because, on the one hand, it is a strategy to mask the actual lability of behavior and to overcome the problems we mentioned above about the difficulties faced by any attempt at classifying behaviors in a robust way; on the other hand, it implicitly makes behavioral homology dependent on structural homology, with the more than likely consequence that the identities established at the behavioral level may turn out to be translatable to identities at the structural level,[10] which, as we pointed out, is not that different from Hodos's original idea of explicitly making behavioral homology always dependent on structural homology.

The idea in the end appears to have turned out to be self-defeating, since it seems to drive us to the paradox that by accepting the historical notion of homology and

---

[10] This is perhaps what is implied by Greene (1994: 378) when he writes: "Behaviors represent very short-term neuromuscular, neurohormonal, and integrative responses of organisms to internal and external stimuli; useful descriptions of behavior can extend necessarily to various external factors as well [...]. However defined, the distinctions between behavior and other phenotypic attributes are fuzzy. Enzyme-substrate reactions and other fast molecular events are referred to as biochemical or physiological, somewhat slower responses are called behavior, and features that appear stable over long periods are known as morphology," which is tantamount to recognizing that the adequate level of analysis is not that of observable behaviors but, rather, some other, deeper level; we come back to this in the next chapter.

behavioral homology made possible by the former, we can only assert the evolution of behavior to the same extent as we can assert the evolution of structure. This leads us to the conclusion that either we should abandon the historical notion of homology or, as already suggested by Klopfer (1969, 1973a, 1973b, 1974),[11] accept that behaviors/functions do not evolve and that the comparative method at this level cannot be used to draw historical inferences[12]—or both, which is, to anticipate one of our conclusions, what we would like to suggest here.

As we had anticipated, this problem has its roots in the assumption that behavior can be studied by making reference to observed behaviors only. All attempts to establish a valid methodology notwithstanding, however, it still remains to be shown that a robust method can be found to determine what counts as a valid and consistent description of a behavior, which is one of the weak points of traditional behaviorist psychology. Thus, as Fodor (1968: 52–53, emphasis in the original) wrote:

> But while it is clear that certain locutions must be denied the status of descriptions of behavior if behaviorism is to be worth discussing, a considerable number of positions are left open as answers to the question of what sorts of descriptions *are* to count as behavioral—for example, descriptions of the states of the musculature of an organism, or of its movements, or of its sequence of postures. As with interpretations of 'logical connection', the possibilities are limited only by the ingenuity of the behaviorist and by the requirement that he take no liberties that would lead to the trivializing of his position.

It is important to stress this point, because underlying the assumption that behavior can be studied just on the grounds of what is observable is the *semantic thesis* that an adequate descriptive vocabulary can be devised to refer unambiguously to kinds or types of observable behaviors.[13] This is relevant for homology as it implies that if two behaviors are homologous, then the terms we use to designate them are

---

[11] We believe that Klopfer (1973) came very close to a similar argument when he wrote that "the important distinction between analogies and homologies is largely dependent on our prior knowledge of the phyletic relationship and physiology of the animals in question. If we already know bats to be mammals and flies insects, then we can classify their wings as analogous structures, a convergent response to a common problem." (Klopfer 1973: 114–115). Note, in passing, that Klopfer's observation is applicable to historical homology in general, not just to historical behavioral homology.

[12] Klopfer (1973), in the immediately preceding paragraph to the one cited in the previous note, seems to suggest something along these lines, namely that behavioral comparisons (and functional comparisons in general) may be useful in drawing inferences on the ecology of organisms, not their history; also, in Klopfer (1974: 215), the author closes his chapter on the evolution of behavior with the following words: "Thus comparative studies of behavior are of themselves unlikely to elucidate the evolutionary sequence or to shed light on the significance of specific differences. Our studies must rather be molar in nature, encompassing both the behavior and the ecology of species that interest us." We believe this idea also permeates the work of Hansell (2005). But all this was already captured by Owen's definition of analogy, which immediately suggests a line of attack against the historical notion of homology: It blurs the difference between homology and analogy.

[13] In this connection, Chomsky's (1959a) devastating analysis of Skinner's (1957) purportedly objective vocabulary to describe verbal (and other) behavior is particularly illustrative.

TABLE 4.1. Guess what these behaviors are for

Behavior	Reference
Feeding a conspecific adult	Nisbet (1973)
Territoriality	O'Donald (1963); Wynne-Edwards (1962)
Pheromone emission	Thornhill (1992); Moore (1998)
Construction of artifacts or shelters	McKaye (1979); Christy (1988); Borgia (1986)
Exhibition of some part of the body	Petrie et al. (1991)
Ritual dance or lekking	Gibson et al. (1991)

These behaviors are all supposed to fulfill the same function and are observed in a wide spectrum of animal species, both within the vertebrates and the invertebrates.

necessarily synonymous.[14] That this program cannot possibly be carried out was one of the central topics of Chapter 3, but we would like to insist on this point here by considering again the behaviors listed in Table 4.1.

If the reader consults the references provided in the right-hand column, she will find out that they all constitute instances of "courtship behavior." But what prevents us from classifying "feeding a conspecific adult" as "altruistic behavior," for example in humans? Or take the case of bower construction by bowerbirds: What are we to make of the (often collective) behavior observed in juvenile bowerbirds whose bowers are not for courtship? Is this perhaps a case of "practicing behavior"? (Hansell 2000: § 8.7, 2005: §5.5). Similarly with pheromone emission, which may be a courtship function in certain insects and aggressive behavior in others (e.g., in beetles; Peschke et al. 1996). Examples could be multiplied at will for these and the other items on the list. The fact is that no correlation can be established between a behavior characterized in the vocabulary of ordinary language and its function, since the latter may be instantiated by many different kinds of behaviors in different species and the same behavior may have different functions in different species. This is the reason why a vast area of studies in the field of animal behavior, in particular those developed within the frameworks of sociobiology and behavioral ecology, although relying on this sort of descriptive vocabulary, insist on the fact that the use of "catchy descriptive labels for various behaviour patterns" (Krebs and Davies 1987: 3) is a mere rhetorical strategy to develop the functional explanations of behavior, which constitute the real focus of their research (Krebs and Davies 1987: Ch. 1). This constitutes a de facto reductive strategy of behavioral descriptions to the purportedly technical functional

---

[14] Owen was well aware of this problem and not in vain did he devote the first pages of his *On the Archetype and Homologies of the Vertebrate Skeleton* (Owen 1848) to discuss such semantic and terminological issues.

language into which hypotheses and explanations are framed;[15] we will immediately come back to the nature of this reduction. Interestingly enough questions about homology have never actually been raised within this tradition—the topic is not mentioned, even in passing, in such works as Wilson (1975), Dawkins (1982), or Krebs and Davies (1987), although it is explicitly committed to explain "the evolution of adaptive behavior in relation to ecological circumstances" (Krebs and Davies 1987: 11). In fact, that homology is not an issue for these disciplines is suggested by John Maynard Smith's intervention in the discussion that follows the paper by Greene (1999):

> The behavioral ecology movement, which developed in the 1960s and 1970s, attempted to analyse the adaptive significance of selective forces operating on behavior [...]. We probably reacted a little too strongly against the fascination that people like Konrad Lorenz had in the homology of different behaviors. Not that I ever intellectually rejected the notion that it is possible to homologize behaviors between different animals, but rather I wanted to be involved with something more testable and analytical. (in Bock and Cardew 1999: 183)

We believe this position is contradictory. Because, if one accepts the idea that behavior evolves and that behavioral characters have the same phenotypic status as morphological characters and, moreover, one is committed to a historical notion of homology (as appears to be the case with most neo-Darwinists), then necessarily behavioral homologies need to be recognized at some level of analysis. It is not our job here to speculate what this level should be, but a quick review of the literature on the topic reveals deep confusion on these matters. Thus, when John Maynard Smith originally defined the notion of an Evolutionary Stable Strategy (ESS), he explicitly stated that "[a] 'strategy' for a contestant is a set of rules" (Maynard Smith and Price 1973: 16), a definition that he refined in a later paper: "A 'strategy' is a specification of what an animal will do in all the situations it may find itself in" (Maynard Smith and Parker 1976: 159; see also Maynard Smith 1974). Moreover, given the assumption that "different behaviours or strategies have different genotypes" (Maynard Smith and Parker: 160), it follows that those genotypes will be selected that underlie strategies capable of enhancing the animal's fitness. Thus, ESSs will evolve following the

---

[15] That this is a kind of reductionism has been explicitly stated by Alexander (1987: 15–16): "[T]he 'How come?' or ultimate cause approach seeks to simplify, explain, or 'reduce' complex phenomena such as human activities by hypothesizing directly that they are parts of even grander sets or combinations of actions. Thus, merely to identify a complex set of activities by a hungry human as an effort to eat is a step in this kind of reduction; another step is to see that effort as part of lifelong effort to build (as a juvenile) and maintain (as an adult) a soma (phenotype, self) in the further interest of using it in the reproduction of one's genetic materials. This reduction by generalizing, because it seeks the adaptive significance of acts, also tends to identify the larger contexts within which specific acts or functions are carried out, and as a result identifies likelihoods of compromise or adjustment in the act that would not be obvious from any other approach."

standard neo-Darwinist model.[16] Leaving aside now the appropriateness of this view, what concerns us here is the ontological status of ESSs. Note that in the two quotes above, there appear the words "rules," in the first, and "specification," in the second one. This suggests that the intended level of analysis is not that of manifest behavior but rather some other more abstract level. In fact, this is for instance the interpretation implicitly assumed by Brockmann et al. (1979) in their study of nesting strategies in digger wasps, and explicitly advocated by Dawkins (1982: 118): "A program (or strategy) is a recipe for action, a set of notional instructions that an animal seems to be 'obeying', just as a computer obeys its program." At first blush it appears that we are dealing here with some kind of cognitive or psychological level that is the actual proximate cause of manifest behavior and, in this sense, behavioral ecology would, in the end, not be far from the position we will be endorsing here: Homologies could be recognized at this abstract level of analysis on the basis of the structural similarities of the strategies, independently of the nature of the observed behaviors caused by them. Unfortunately, *pace* Dawkins's claims to the contrary, it is far from clear that such an interpretation is possible for strategies, which are more often than not just mere descriptions of the manifest behavior of animals in specific situations, a fact explicitly recognized by Krebs and Davies (1987: 222) in their discussion of the issue:

> Some authors prefer to use the word 'strategy' to give a complete specification of what an animal will do when competing for a scarce resource, and the word 'tactic' for the behavioral components of a strategy. [...] Although this distinction is fine in theory, as we shall see it is often difficult to distinguish tactics from strategies and we prefer to use the word strategy somewhat loosely to describe any behavior pattern or structure used by an individual to compete for a scarce resource.

It is these kinds of inconsistencies that motivated the strong criticisms of behavioral ecology and sociobiology by evolutionary psychologists (Cosmides and Tooby 1987; Tooby and DeVore 1987) and which we share almost entirely, despite the fact, of course, of not sharing evolutionary psychology's commitment to the adaptationist program.

Turning now to the other tradition in the study of animal behavior, namely the one more closely linked to the ethology of Tinbergen and Lorenz, we see that, in this case, homology *is* an issue. It always has been. As Konrad Lorenz wrote: "A great part of my life's work has consisted in tracing the phylogeny of behavior by disentangling the effects of homology and of parallel evolution" (Lorenz 1974: 231). This being a main concern in this tradition it is interesting to note that, unlike sociobiology and behavioral ecology, it always aimed at a rather different kind of reduction in order

---

[16] Actually, the model is a little more complex, since it is assumed that genotypes will also independently determine what role among the possible ones the animal will play possessing a specific strategy (Maynard Smith and Parker 1976), but this refinement is irrelevant for the point we want to make in the text.

to describe behavior, namely one based on form or structure and, crucially, one that could eventually fill the gap between behavior and neurobiology—see Kennedy (1992: §3.7) for discussion. Thus, for example, be it through Fixed Action Patterns (FAPs; Tinbergen 1951), Modal Action Patterns (MAPs; Barlow 1977), or, more recently, the Eshkol-Wachman notation (EW; Golani 1992), ethologists have strived to find a simple, objective, and hopefully universal language to characterize the movements associated with some piece of behavior and caused by specific neurophysiological mechanisms. It is symptomatic, however, that both FAPs and MAPs were already defined as stereotyped patterns *within* species and that EW, although often presented as potentially universal, has been criticized precisely on the grounds that it appears to work well with a limited set of related species but fails with other, more distantly related ones, even within vertebrates (Byers 1992). This takes us back to our previous discussion in the context of Greene's (1994) defense of the concept of behavioral homology: By steering away from functional considerations and coming closer to form one eventually ends up blurring the difference between behavior and morphology and making behavioral homologies entirely dependent on structural ones; but this also makes them trivial, since they work within limited sets of species and one can hardly ever extrapolate them over others.[17] Moreover, it is questionable that any historical sense of homology is actually applicable here and, in fact, Golani (1992: 299) concludes that "a definition [of homology] based on common descent precludes conclusions about descent from judgments of behavioral homology." A consequence that Greene (1994) appears to have failed to appreciate.

It seems to us that *all* biology just failed to see these consequences when it invented "behavioral homology." Indeed, it failed to see that the fuzziness of behavior was an accepted fact by behaviorists, that all this followed directly from Wittgenstein's (1953: §66ff.) notion of "family resemblance" and that this was precisely the reason why behaviorism always tried to draw a distinction between *central* cases of a behavior versus *peripheral* ones.[18] Thus, biology just uncritically imported the idea of both Ryle (1949) and Wittgenstein (1953) that "[f]or each mental predicate that can be employed in a psychological explanation, there must be at least one description of behavior to which it bears a logical connection" (Fodor 1968: 51) and epitomized in Wittgenstein's aphorism "Ein 'innerer Vorgang' bedarf äußerer Kriterien" ("An 'inner process' stands in need of outward criteria;" Wittgenstein 1953: §580).[19] But this

---

[17] The alternative of reducing to the limited vocabulary of "respondents," "operands," "reinforcers," etc., as suggested by Skinner (1938, 1957), appears to be even less viable and liable to produce interesting extrapolations to other behaviors than those concerning the experimental situations for which these terms were devised; see Chomsky (1959a).

[18] See Armstrong (1968: Ch. 5) for a detailed discussion and criticism of this point.

[19] Both Ryle and Wittgenstein denied that they were behaviorists, but, as pointed out by Armstrong (1968: 55), "the only reason that these philosophers denied that they were Behaviourists was that they took Behaviourism to be the doctrine that there are no such things as minds. Since they did not want to deny the

uncritical import of the behaviorist assumption that psychological vocabulary can be reduced to descriptions of observable behaviors and that those predicates for which no such reduction was possible had to be banned from psychological explanations did not come alone. This eventually takes us to assumptions 2 and 3 in our list above, which, as we would like to show, are also an integral part of the contemporary biological approach to the study of behavior.

Let us turn then to dispositions.

4.1.2 What are the causes of behavior?

Gilbert Ryle defined dispositional properties as follows (1949: 43):

A statement ascribing a dispositional property to a thing has much, though not everything, in common with a statement subsuming the thing under a law. To possess a dispositional property is not to be in a particular state, or to undergo a particular change; it is to be bound or liable to be in a particular state, or to undergo a particular change, when a particular condition is realised.

Which immediately takes us to the general schema for psychological explanation assumed by behaviorists and illustrated by Ryle (1949: 50) with his example of the broken glass: "The explanation is not of the type 'the glass broke because a stone hit it', but more nearly of the different type 'the glass broke when the stone hit it, because it was brittle'." This, translated into behavioral talk would look more or less as follows: "The snake showed a constriction behavior when it saw the prey, because it possesses the disposition to constrain." The Panglossian flavor of this explanation is surely not accidental; and it is perhaps no coincidence either that biology, implicitly or explicitly, has accepted these kinds of explanations wholesale, which square perfectly with traditional adaptationist explanations (Gould and Lewontin 1979). In fact, that the connection exists is clear from Dennett's (1987: Ch. 7) claim that the so-called "Panglossian Paradigm" is a natural extension of his Intentional System Theory (IST) and of his acknowledging that IST is a "sort of holistic logical behaviorism" (Dennett, 1987: Ch. 3), but it is perhaps better appreciated in Millikan's (1984, 1993) theory of "proper functions" and her idea that "it is the 'proper function' of a thing that puts it in a biological category, and this has to do not with its powers but with its history" (Millikan, 1984: 17). Thus, just as organisms act as they do because they have been subjected to certain histories of reinforcement, certain functions/behaviors have been preserved because they have been subjected to certain histories of (natural) selection.[20]

---

existence of minds, but simply wanted to give an account of mind in terms of behaviour, they denied that they were Behaviourists." See also Chihara and Fodor (1965).

[20] A connection, by the way, that Skinner himself explored in Skinner (1953, 1975) and is explicitly assumed by Dennett (1995, 1996); see also Amundson (2006).

The question of predispositions is important because it concerns what Krebs and Davies (1987) would call the causal explanations of a behavior or what in more traditional views constitute the control mechanisms of behavior (e.g., Klopfer 1974: Ch. 11). Here the causal story is intricate, since different types of causality are involved. First, we have the question of what causes which behavior in a specific situation; this is, in principle, what is meant by "causal explanation" or "control mechanism" of a behavior. But, second, questions at this level of causality are intimately related to questions at the ontogenetic or developmental level of causality, i.e., with answers to the question how a behavior develops in the individual (e.g., questions about learning, our point 3 above). Third, questions about ontogeny eventually lead us to questions on evolution and history, which concern the causes of the preservation of some specific behavioral character.

Let us see how all these causal stories are connected.

The central point in the first level of causality is the assumption that behavior is always stimulus-driven, where "stimulus" is in general some kind of external cue that triggers some kind of neurophysiological response in the organism. We would like to emphasize that in this model there is no room for genuine psychological explanations in the sense that no intermediate level is assumed to exist between the stimulus-generating environment and the response-generating neurophysiology.[21] This model is constructed on an analogy of cybernetic models based on control theory (Klopfer 1974: Ch. 11; Kennedy 1992: §4.2) where the response of a device is controlled by a series of stimulus detectors in such a way that the device "behaves" *as if* it had purposes or goals.[22] The "as if" part is relevant here, since it is precisely the intentional vocabulary of purposes, beliefs, desires, etc., that logical behaviorism considered to be eliminable in all explanations of behavior. The cybernetic analogy breaks down however when no external cue seems to be available to act as the cause of some behavior. In these cases, the strategy is to keep the stimulus-response model by assuming either that some responses are merely "displaced" or "delayed" with respect to the time in which the external stimulus occurred or that the organism

---

[21] It thus constitutes a kind of eliminative reductionism where "mental talk" is to be avoided in favor of a vocabulary whose terms refer only to observables, i.e., behavioral events and/or neurophysiological events. As pointed out by Fodor (1968), however, this position is only understandable under a strict reading of Ryle's (1949) dismissal of Cartesian dualism on the assumption that "mental talk" is incompatible with materialism and, therefore, that the only real alternatives to dualism are either some form of behaviorism or some form of the Central State Theory of Place (1956), Smart (1959), and Armstrong (1968). What this position fails to see is that "mentalism" is not incompatible with materialism or physicalism, that, as argued in detail by Fodor (1968, 1975), there is a third way, and perhaps the only one, by the way. Ironically many of those researchers who have adopted physicalist positions in the study of animal behavior and rejected mentalism as a form of anthropomorphism have ended up being more Cartesian than most neo-Cartesians like Chomsky or Fodor and accepting the idea that an unbridgeable divide exists between animals and humans; see Kennedy (1992) for a representative example of the position that only humans have minds.

[22] A number of illustrative, and ingenious, examples of these kinds of cybernetic explanations can be found in Braitenberg (1984).

possesses some kind of internal stimulus-generating mechanism capable of triggering the observed behavior without the presence of an external cue. In this latter case ethologists usually talk about "drives" (another word for "predispositions") and thus we find expressions such as the "drive to mate" or the "drive to eat," which are often explained as internal stimuli generated in order to restore the homeostatic equilibrium of the organism. Note how this model fits perfectly into the dispositional mode of explanation proposed by Ryle (1949). Thus after the observation of, say, a snake catching a prey we would say that "the snake caught a prey because it had the drive to eat," where "drive" here stands for the set of neurophysiological events triggered in order to restore the homeostatic equilibrium, previously altered by some other external event, such as the deprivation of food. The problem with these kinds of explanations is that, whereas they may work with relatively simple behaviors, they face insurmountable problems with more complex ones, nest building being a case in point. Compare the case of the snake catching a prey with "the bird built a nest because it had the drive to build a shelter," where it is hard to find a convincing neurophysiological story, free of "mental" vocabulary, to explain the complex process of building a nest. Note, moreover, that appealing to "genetic predispositions" (Krebs and Davies 1987; Wilson 1998: Ch. 7) hardly changes the picture very much.

The nest building example has the virtue of showing the extent to which dispositional analysis is vacuous. The problem is that predispositions lack any causal powers entirely, or as Fodor (2008: 49; emphasis in the original) puts it, "*mere dispositions don't make anything happen*. What causes a fragile glass to break isn't its being fragile; a glass that is fragile may sit intact on the mantelpiece forever. What causes a fragile glass to break is *its being dropped*."[23] Thus, inasmuch as the traditional Cartesian account of the mental (internal) causes of behavior is rejected, the only possible alternative is to resort to some external cause to explain the occurrence of some piece of behavior. This, in principle, works for the simplest cases, but fails with the more complex ones like nest building where no such external cause is patently present. In these latter cases the only possibility is to appeal to the disposition (or drive) itself, but what makes the disposition "active"? Of course, the only possible answer is either instinct or learning (or a combination of both). In this case it is common practice to establish hierarchies of organisms in terms of their learning capabilities, as in Bateson's (1972) "zero learning" to "learning III" typology, Wilson's (1975) "instinct-reflex machine", "directed learner", and "generalized learner" hierarchy, or Dennett's (1995, 1996) "Skinnerian", "Popperian", and "Gregorian" creatures. Here questions of

---

[23] Or, in Goodman's (1954: 41) words: "To find non-dispositional, or manifest, predicates of things we must turn to those describing events—predicates like 'bends', 'breaks', 'burns', 'dissolves', 'looks orange', or 'tests square'. To apply such a predicate is to say that something specific actually happens with respect to the thing in question; while to apply a dispositional predicate is to speak only of what can happen."

ontogeny mesh up with questions of phylogeny, but this is expected in this context, since as we have already pointed out in note 17 and emphasized by Dennett (1995: 374), "B. F. Skinner was fond of pointing out [that] operant conditioning is not just analogous to Darwinian natural selection; it is continuous with it." We thus have a *scala naturae* of increasing plasticity with simple reflex-driven organisms at the bottom and organisms capable of somehow monitoring their learning at the top. This idea of a progression from specific learners to generalized learners does not, however, make any sense even from the point of view of the most orthodox adaptationism. As clearly illustrated by Cosmides and Tooby (1987: 294–299): "[T]here is no domain general criterion of fitness that could guide an equipotential learning process towards the correct set of fit responses."[24]

4.1.3 Conclusions to this section

Our main concern in this section has been to pinpoint the problems derived from the application of the comparative method at the functional/behavioral level in order to infer homological relations. As was also the case in the previous chapter, we've seen that the different strategies to homologize functions/behaviors fail either because the functional/behavioral categories applied in the descriptions remain ungrounded or because their grounding on a solid structural basis actually has the consequence of blurring the functional/behavioral category to the extent of making it virtually indistinguishable from the structural category on which it has been grounded. The main conclusion, which is already becoming recurrent in this work, is that from the observation and analysis of behavior one can derive insightful generalizations on its structural/formal basis that make any functional considerations superfluous.

The ethological tradition initiated by Konrad Lorenz and Niko Tinbergen came close to this conclusion with its cybernetic model to account for the causes of behavior, but failed because it was unable to recognize an intermediate structural level of analysis between observable behavior and neurophysiological activity. As we have seen, the vocabulary of neurophysiology is unable to account in a satisfactory way for those behaviors that appear to involve higher cognitive capacities—instead of simple neurophysiological responses to physical stimuli—like, for example, nest building. In our opinion, the only possible way out of this blind alley is to recognize that an additional level of structural organization exists on top of neurophysiology, subserving such complex behaviors: The computational level. The rest of this book is devoted to showing how a biologically sound model of this computational level can be developed and how it can be used to elaborate sound biological hypothesis on the

---

[24] Incidentally, generalized learning (aka learning by induction) has been known to be problematic at least since the time of Hume (hence the label Hume's Problem). Goodman (1954) offered a solid philosophical argument showing that it is in fact impossible; more recently Wolpert and Macready (1997) proved it mathematically (also Ho and Pepyne 2002).

structure, the development and the evolution of cognitive capacities. The notion of homology will play a central role in the construction of our proposals and, consequently, we believe that we cannot put an end to this chapter before turning to homology and its many definitions, in order to underpin our proposals on much more solid grounds.

4.2 Why a Historical (Darwinian) Concept of Homology Doesn't Work—Or Why We Are in Need of an Ahistorical (Owenian) Concept of Homology

As we saw in the previous section, the failure of the idea of "behavioral homology" is twofold, in that it is completely dependent on homological relations previously established on either structural or historical grounds. There we argued at length for the priority of structure over behavior. It is time now to focus our argument on the priority of structure over history. If our case is on the right track, then the only conceptually acceptable notion of homology is a structural and ahistorical one, i.e., the concept as originally used by the classical morphologists—Russell (1916) for an overview—and rightly defined and formalized by Richard Owen (1843, 1848, 1849).

Owen thought of "homology" as a form of "sameness" between organs due to a shared internal quality, which he named the "archetype" (Owen 1843, 1848, 1849). His idea thus departed from both the unrestricted form of identity that Geoffroy Saint-Hilaire founded on the observation of common patterns of correlation between the components of the parts of different organisms (Geoffroy 1818, 1830) and the form of identity that Aristotle in his *Parts of Animals* restricted to the organs serving an identical function in different animals, for which both authors employed the term "analogy." For Owen, as it is well known, structures could be homologized in spite of their serving very different functions (swimming, flying, digging, etc., as in the case of homologue vertebrate forelimbs; Owen 1849). Furthermore, and apparently not as well known, a similar pattern of correlation was not enough for Owen to homologize structures, as shown by the fact that he opposed Geoffroy's contention that vertebrate limbs and invertebrate appendages were homologous structures (Geoffroy 1820; Owen 1849). Something more was required to declare two or more structures homologues, an "internal quality," as we have said, causally acting on the differentiation of these structures within the organic whole and whose specific pattern of development in each organism was responsible for the superficial diversity of organic forms and functions against a background of deep natural identity (Balari and Lorenzo 2012).[25]

---

[25] For the sake of simplicity, but also for its secondary role in our argument, we will put aside "serial homology" (i.e., repetitions of the same part within a single organism) from our discussion. It is commonly accepted, however, that serial homology fits more comfortably within ahistorical approaches to homology.

For Owen, archetypes were organism-internal qualities constraining the development of forms to certain patterns of organization—not Platonic ideas to which actual natural forms should accommodate in some mysterious way, contrary to accepted common wisdom (Rupke 1994; Balari and Lorenzo 2012).[26] If, as seems to be the case, Darwin was unconvinced by the Owenian concept of homology because it dangerously invited a Platonic vision of nature, we must conclude that his reading of his contemporary was based on an unfortunate misunderstanding and that his own historical concept of homology relies on a false start. Thus, the numerous problems associated with this latter concept, which we discuss below, do not come as a surprise.

The history is, we think, well known (see Amundson 2005: Ch. 4). While trying to avoid the ghost of the archetype without losing the opportunity of explaining the "unity of type" to which Nature appears to obey, Darwin concluded that such a unity was nothing but heritage directly coming from ancestors, more evident insofar as we are dealing with closely related lineages of organisms and much fuzzier when dealing with loosely related ones (Darwin 1859: Ch. VI). There was no need of archetypes, Darwin thus concluded, in order for the traces of ancestors to be present in their descendants: Homology was nothing but a form of identity due to common ancestry. He was confident that with this move abstract types were definitely barred from biological theorizing in order to explain how the parts of different organisms correlate with each other. However, we rather think that what he actually did was to extract from the inside of organisms the foundations of their identity and to locate it in an entity external to them. Darwin thus transformed homology into a form of identity due to the existence of a shared "external entity" (an "ancestor") (Simpson 1961; Mayr 1982), instead of, as in Owen, a shared "internal quality" (an "archetype"): In a way, a style of thinking much closer to Platonism than Owen actually ever practiced.

On the problems of serial homology to the historical perspective, see de Beer (1971), Van Valen (1982), or Ghiselin (2005).

[26] We invite the reader to compare the two following quotations and to judge for herself the modernity of Owen's ideas:

Nevertheless, I am constrained by evidence to affirm that in the vertebrate as in the invertebrate series there is manifested a principle of development through polar relations, working by repetition of act and by multiplication of like parts, controlled by an opposite tendency to diversify the construction and enrich it with all possible forms, proportions, and modifications of parts, conductive to the fulfilment of a preordained purpose and final aim: these opposite yet complemental factors co-operating to the ultimate result, with different degrees of disturbance, yet without destruction, of the evidences of the typical unity. (Owen 1866: x–xi)

In the development of these respiratory structures, a complex genetic control has been superimposed on the simple iterativity of a self-similar process that could produce per se the fractal geometry of the organ. (Minelli 2003: 52).

And, as a matter of fact, the historical (or Darwinian) concept of homology goes together with its own ghost: The ghost of circularity (Boyden 1947; Rieppel 1992, 1994; Brigandt 2002). On the one hand, the homology of parts is seen as proof that the organisms showing them are descendants of a common ancestor that already showed this very same part; but, on the other hand, the existence of a common ancestor is seen as the proof that the parts of different organisms can be safely seen as results of the diversification of a single part (i.e., as homologues), instead of independently evolved answers to similar environmental pressures or some other hazardous cause (i.e., analogous or homoplasic forms). So, the proof of what needs to be proved needs in turn to be seen as proof of what needs to be proved. If anything, this is a circle.

However, the classificatory principles of cladistics (Henning 1966) came to the rescue of the historical concept of homology, by augmenting its definition with a crucial additional adverb: Similar parts of two or more organisms are true homologues if and only if it can be *independently* settled that these organisms are descendants of the same ancestor; or, using cladistic jargon, if it can be independently settled that they exhaustively compound a unique taxon (Stevens 1984; Lauder 1994; Rieppel 1992, 1994). Thus, testing a homology requires: (i) Taking into consideration a certain meaningful number of characters, with the exclusion of that whose identity is being scrutinized (x in figure 4.1 below); and (ii) plotting into a tree topology, in the most economically conceivable way, the family relations of the species so compared (figure 4.1). Homology will be concluded to exist if possession of a similar (thus, identical) character x replicates the continuity of a taxon (as in the descendants of $A2$ in figure 4.1); otherwise, if similarity is interrupted by a phyletic discontinuity (as in the descendant of $A3$ in figure 4.1), it will be dubbed a homoplasy.

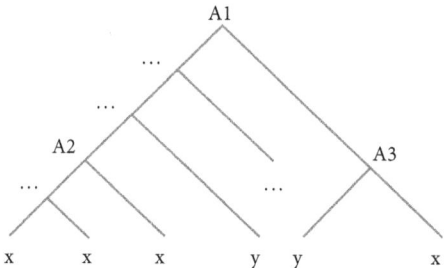

FIGURE 4.1 Guess which xs are the same x (or a different x)

Cladistic analysis imposes strong constraints on the identification of homologies. Let us suppose x = have wings. Then all xs present in the descendants of $A2$ are homologues because there is a continuous line of ancestry starting with $A2$ (which will be inferred to also possess x). The x in the descendant of $A3$ cannot be classified as homologous because no continuous link exists with the other xs and will therefore be considered to be a case of convergent evolution (homoplasy), as is the case with the wings of bats and birds, for example.

This methodological maneuver, however, does not rescue the historical concept of homology from another, even more fundamental, problem of conceptual circularity. Homology names a form of "identity" (namely, identity of parts of different organisms) and, therefore, a symmetrical relation between any two characters so connected. This purports to say that if two or more organisms are made up of homologous structures, each of the structures at hand is a "homologue" in exactly the same way and to the same extent as the others. The historical concept of homology, however, relies on the attribution of a privileged or special status on the ancestor's part, relatively to the corresponding parts of the descendants. As a matter of fact, in order for the historical concept to be operative, a different kind of relation must be supposed to be held between the descendants than between any of the descendants and the ancestor. However, according to the same concept, the descendants' parts can only be deemed true homologues if they are modified variants and thus the same part of the ancestor they all come from. Therefore, the ancestor's part must be a "non-homologue" of the descendants' parts for the historical concept to work, but at the same time it must be a "homologue" of them all for the concept of homology to be applicable to begin with.

Summing up, the historical concept of homology requires a kind of asymmetrical relation between the ancestor's part, on the one hand, and the parts of the descendants, on the other hand, so as to avoid circularity (figure 4.2, right). This asymmetry leads to a contradiction, though: If the ancestor's part is not a homologue of those of the descendants, the latter cannot be modified variants of the same character and, therefore, true homologues of each other. Homology entails symmetry without exclusion of the ancestral character (figure 4.2, left), or it is not true homology.

FIGURE 4.2 **Guess whether the ancestor's part is or is not a homologue**
Darwin defined evolution as "descent with modification," and, at the same time, assumed that two characters in two species were homologous if, and only if, they were inherited from a common ancestor. Thus, the characters labeled $P1$ in $D1$ and $D2$ are homologous because they can be traced back to $P1$ in A (left). The problem is that "homologous to" is a symmetrical relation, therefore $P1$ in A is just as homologous to $P1$ in $D1$ and $D2$ as $P1$ in $D1$ and $P1$ in $D2$ are homologous to each other (right); but $P1$ in A cannot be, for example, homologous to $P1$ in $D1$ according to Darwin's definition, because A and $D1$ do not have a common ancestor, since "ancestor of"—unless we assume that it is reflexive (and A is considered to be an ancestor of itself), which doesn't make much sense—is an asymmetrical relation.

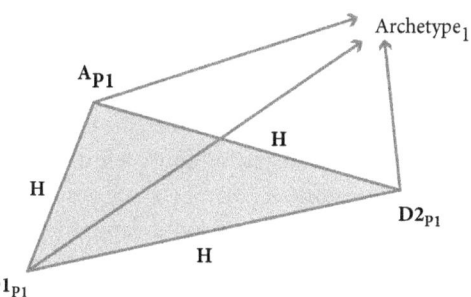

FIGURE 4.3 Who's afraid of the archetype?
The relation of special homology, being symmetrical, defines what in graph theory is known as a complete graph, where every pair of vertices is connected by a single edge (shaded triangle), unlike the relation "ancestor of," which defines a tree. The archetype adds an additional dimension to homology (general homology) and a frame of reference against which special homologies can be validated.

Consequently, homology cannot be founded on ancestry, because ancestors take part in the relation to the same extent as the descendants.[27]

Our conclusion is thus that homology is not to be fully explained (cannot be fully explained) by common ancestry if ancestors also partake in the relation (as they actually do), and that the alternative is restoring Owen's original concept of homology as based on an "inner quality" shared by both ancestors and descendants. This amounts to saying that the reinstatement of an ahistorical concept of homology is in order, as well as the rehabilitation of a biologically acceptable concept of the archetype (figure 4.3 and Young 1993). All this does not purport to deny that homological relations can be inherited and thus partially explained on historical grounds, of course; it just boils down to the claim that "homology" is not a historical concept and—heuristic considerations apart—it stands in need of ahistorical criteria.

Actually, this fresh start of the question calls for both Owenian concepts of "general homology" (identity given an archetypal background) and "special homology" (identity between different species' parts, independently of their historical relations) (Owen 1848), equally unavoidable on logical grounds. If the opposing comment is that, with

---

[27] A similar argument has been put forward by Brigandt (2002), who contends that the historical approach leaves the nature of the relation between the part of the ancestor and the different homologues in its descendants unexplained. In fact, it seems that an important source of complication for the historical notion of homology is that the relation "being a modified descendant of" is asymmetric as it inherently implies the idea of transformation and difference, whereas homology necessarily relies on some idea of sameness, but historical homology appears to collapse both relations into one—a contradiction. Moreover, historical homology reintroduces the vexing question of the origins of a trait (Minelli 2009: Ch. 1), which easily leads to an infinite regress. Note that as soon as history is left out of the equation, and homology is established on ahistorical criteria, the problem of origins disappears. We are grateful to Alessandro Minelli for fruitful discussions on this issue.

this move, the comparative approach becomes essentialist (Wake 2003), the answer is "yes, necessarily so." But this is not really at pains with the normal practice of biology, because "essences," as required by the argument, are nothing but the determinants of form in the very sense that biologists routinely seek them.[28]

Ghiselin's (2005) variant of the historical approach deserves, however, a special comment. He contends that homology names a "correspondence" relation—Van Valen (1982) and Brigandt (2002) also insist on this point—and not, as usually and sometimes unconsciously assumed, a "similarity" relation—see West-Eberhard (2003: Ch. 25) for an explicit characterization of the concept along these lines. The difference is crucial for Ghiselin, because "correspondence" relates entities irrespective of the properties of each one, while "similarity" entails the existence of a certain cluster of properties shared to a certain degree. From this, Ghiselin concludes that the only criterion capable of constraining the application of the concept of homology is the existence of continuity in terms of historical kinship. The impact of Ghiselin's definition on the historical concept of homology is that the criterion of "common ancestry," which problematically excludes the ancestors, is replaced by a criterion of "continuity of kinship," which applies to ancestors and descendants to the same extent. Thus, the contradiction that follows for not treating ancestral parts as homologues to the corresponding parts of the descendants is overcome. The price of the strategy is however high (and not worth paying), because it virtually reduces the concept of homology to no meaning at all and makes it biologically inoperative. A meaningful concept of homology refers to relations of "transformed identity"— "The same organ under every variety of form and function;" Owen (1843: 374)—and it brings together a natural class of entities subjected to lawful processes having to do with the evolution of development (Wagner 1996; Brigandt 2002, 2009). However, a bare relation of "correspondence," as Ghiselin himself acknowledges, brings together entities—parts of organisms to which a familiar bond is credited—but for no particular biological reason at all. Therefore, we think that it does not qualify as homology in any technical sense.

On the contrary, an advantageous consequence of abolishing the historical concept of homology is that it gives us back the opportunity of establishing homological relations in spite of evolutionary distance and even if phyletic intermittences exist between the organisms being compared—for example, between the descendants of $A2$ and $A3$ in figure 4.1 above. As a matter of fact, the barring of this kind of relations is another anomaly for the historical concept of homology, as observed, for example, by West-Eberhard (2003: Ch. 25). The distance between certain classes of organisms notwithstanding (such as arthropods and vertebrates), the processes leading to the

---

[28] Actually, papers such as Brigandt (2002, 2009) or Wilson et al. (2007) can be read as an update of the essentialist approach in biology. In the actual practice of biology, this philosophical reorientation has been predated by such seminal work as that of Slack et al. (1993) on the definition of "animal."

development of some of their parts (such as appendages) can be deemed deeply identical (Shubin et al. 1997). From the historical point of view, however, these relations are necessarily homoplasies, that is to say, cases of parallel evolution of similar characters (Striedter and Northcutt 1991), independently from the developmental resources acting on their individuation and stabilization. But the consequence of this is that, when faced by these situations, the historical point of view has to overlook the fact that parallelisms can be the reflection of shared "underlying generators" (Gould 2002) or "developmental inertias" (Minelli 2011), opening the door to well established phenomena such as "latent" (de Beer 1971) or "deep" (Shubin et al. 2009) homologies.

Furthermore, the historical concept of homology points to the dubious conclusion that parts can be deemed homologues at a certain level of kinship—for example, wings of "bats" ($A2$; figure 4.1), wings of "birds" ($A3$; figure 4.1), appendages of "arthropoda," limbs of "vertebrata" and so on, but mere homoplasies at other, more general, levels—for example, appendages of "animalia" ($A1$; figure 4.1). However, given that being diversified variants of the same is within the very meaning of homology, applying in addition the concept of homoplasy to these cases introduces an unnecessary element of confusion to the analysis. Note that the result should read as if the historical process had acted transforming into non-identical—and not just diversifying, what, at the same time, is identical, a conceptually clumsy idea yet widely accepted in comparative biology.

Geoffroy was most certainly right (against Owen's opinion) when he contended the deep natural unity of the appendages of animals.[29] But if we are today more persuaded to accept ideas like this than we were some decades ago, justice must be done to the biologically updated and ahistorical—i.e., "Owenian," concept of homology leading to such results.

4.3 Conclusions

The evolutionary study of behavior predates modern evolutionary thought and actually exerted a direct influence on the modeling of Darwin's evolutionary ideas, as Richards (1987) has convincingly argued. This traditional behavioral evolutionism

---

[29] Minelli's (2000, 2003, 2011) intriguing proposal concerning this question is that appendages in different kinds of organisms—such as arthropods and vertebrates—are not descendants from the appendages of a shared ancestor, but a translation to the proximal–distal axis of developmental resources governing the establishment of the main—antero-posterior—axis, independently attained in different animal groups. This phenomenon of "axis paramorphism" introduces, without hesitation, an instance of homology entirely independent of the continuity requirement of the historical concept and is therefore (with qualifications) a variation on the idea of "deep homology;" so, for example, Minelli (2000: 164) concludes: "vertebrate limbs and arthropod limbs are not historical homologs, but homoplastic divergent duplicates (paramorphs) of historical homologs," which, in our opinion, is tantamount to saying they are "ahistorical homologues."

(Erasmus Darwin, Pierre-Jean Cabanis, Jean-Baptiste de Lamarck, among others) had its roots in the sensationalist heritage of Lockean empiricism, according to which an organism's sensitive powers were enough to make sense of different environmental stimuli and to associate them within advantageous patterns of activity transmitted and enriched generation after generation. Things seem not to have changed much with the advent of modern Darwinism, with behavior still a locus of descent and, consequently, of direct historical connections between organisms. We have shown in this chapter that, just as pre-Darwinian evolutionary approaches to behavior relied on sensationalism, modern Darwinian approaches on the matter crucially rely on a behaviorist ideology that, like sensationalism, is not tenable anymore: Minds are not just passive receptors of impressions coming from the outside, but structured entities that react accordingly to environmental stimulation; rather, it is behavior that is so transient, blurrily limited, and multifariously describable that no structure can be assigned to it independently of the mental workings underlying it. So it is on mental structures and operations, instead of observable behaviors, that the focus must be shifted when trying to establish evolutionary relations between organisms.

Furthermore, we have more specifically shown that homologizing behaviors seems only to work given two additional premises: (i) Similar structures underlie the behaviors at hand; or (ii) they are manifested by closely related and, consequently, similarly structured species. Both premises (i) and (ii) show that the notion of behavioral homology is strongly linked to that of structural homology (but not the other way around); premise (ii) further shows that the notion of behavioral homology is also linked to that of historical homology, which we have independently argued that is, under its different guises, circular, contradictory, or vacuous.

So a paradigm shift is urgently needed in order for the concept of homology to be extended to the realm of cognition.

5

Computational Homology

> Thought relates to 'brain' of man as does electricity to the nervous 'battery' of the torpedo: both are forms of force, and the result of action of their respective organs.
>
> Richard Owen, 1868

We have argued at length that the concept of homology is intrinsically structural and, from now on, we will be assuming that "structure" is the key idea for any approach to the evolutionary study of language. Traditionally, biological structures were defined as patterns of correlations among the component parts of organs and, consequently, homologies—though not always receiving this name—were established attending to the identity—or quasi identity—of patterns of correlation observed in different organisms—see Geoffroy (1818). Nowadays, however, structural correlations are considered sometimes deceitful in that similar patterns of organization can nevertheless be the outcomes of very different generative processes, which seriously calls into question that surface similarities are in these cases the expression of a deep natural unity—see Wagner (1989a, 1989b) for some illustrations. As a matter of fact, Owen (1849) himself was already aware of this when he observed that an apparently similar pattern of correlations—as in the case of the limbs of vertebrates and the appendages of invertebrates—was not enough to deem the corresponding systems as homologues. Consequently, he added certain developmental criteria to the concept of homology—such as following parallel developmental pathways from equivalent embryonic locations. This, we think—see Balari and Lorenzo (2012), makes Owen a clear precursor of the modern "biological concept of homology," as understood from Wagner (1989a, 1989b) on, according to which structures are identical—i.e., true homologues—insofar as they share developmental resources and constraints. The connection so established between the concepts of "structure" and "development" will be demonstrated to be crucial in the next chapters of this book.

For the moment, we will concentrate on another concept to which—crucially, for us—the idea of structure also points, namely, the notion of "activity," as established in Love (2007) and already alluded to in Chapter 3, where the importance is stressed of tracing a clear-cut distinction between what organic structures "do"—their "activity"—and what they "are for"—their "use." Love's main point is that functional analyses often mix these very disparate concepts and as a consequence they cannot provide an insightful basis to evolutionary thinking. On the one hand—he contends, the idea of "use" has to do with the actual workings of an organism located in some particular environment in which its component organic systems render it certain practical benefits—say, attracting mates, avoiding predators, exchanging information, and so on. There is, however, a serious problem with this concept of "use," namely, that it is built on the idea of "function/behavior," with all the shortcomings we pointed out in the previous chapter. On the other hand, the concept of "activity" has to do with the operations a certain structure is capable of given the organization of its component parts—say, blood circulation, food processing, antibody generation, and so on. This concept thus relies on an idea of "function/structure," free of all the inconveniences of the concept of "use." In point of fact, the idea of "activity" just captures the dynamic dimension of structures, from which it is ultimately impossible to tell them apart. This purports to saying that activities are directly and inherently outcomes of development—and, given premises we will introduce in the next chapter, also of evolution, which, on the contrary, cannot be said of uses. This is not to deny any role for "functions/behaviors" in evolutionary explanations. In any event, we contend that such a role can only be an indirect and secondary one, for two different reasons: (i) because the activity of a structure is a determinant for the structure's putative uses—but not the other way around; and (ii) because the uses/benefits associated with structures can explain the proliferation of certain organisms—i.e., the bias toward one or another phenotype within a population, but not the origins of these organisms to begin with, which consist of the introduction of new competing phenotypes—i.e., new structures/activities—into populations. This is a position at odds with current adaptationist thinking, and concomitantly our major point of disagreement with contemporary evolutionary psychology, a point we have already stressed before in this book.

With all this in mind, we start this chapter by defending the idea that "computing" is the name of a natural organic activity—namely, a mental activity—carried out by a particular structure of complex nervous systems. This activity "is for" nothing in particular, it just subserves many different abilities within different organic and environmental contexts, and—this practical and epiphenomenal diversity aside—is open to a narrow, but significant, range of inter-specific variation. We then focus our attention on the material and developmental basis of natural systems of computation and, putting all these pieces together, turn to this chapter's main question: How the triad "activity/structure/development" can help to connect the Faculty of Language (FL)—i.e., the human Central Computational Complex (CCC_{HUMAN})—with a reliable base of true homologues and, thus, to unravel its evolutionary origins. Our reasoning will thus describe a trajectory leading from activity to development, with the notion of structure as a pivotal concept all along the way, so fulfilling the project that we started in the previous chapter.

5.1 "Computing" as a Natural Activity (or "Animal Computation" as a Natural Kind)

Our basic contention in this section is that complex nervous systems contain a particular organic structure whose activity is the natural counterpart of that of digital computers. This is not to be read as claiming that nervous systems are mere composites of such structures or that mental activity is, by and large, computational activity. These are matters that lie beyond the scope of our speculations and about which we declare ourselves agnostics. Our much more humble claim is that the brains of complex animals contain (at least) one such system in charge of their symbol manipulation capabilities. Details concerning the physical instantiation of this system will be given in a following section. For the moment, we will concentrate on its activity and we will describe it using an abstract vocabulary.

The basic premises of our argument are: (i) That mental activities are held (at least to a certain extent) on representations; (ii) that mental representations are made of symbols; and (iii) that symbol manipulation is the means whereby mental activity becomes capable of governing intentional and goal-oriented behaviors of organisms.[1] The concept of "computation" refers specifically to (iii), once such an activity is put in connection with a particular cognitive architecture, i.e., to an actual device in which the functional specializations required to run this activity are explicitly settled (Gallistel and King 2009). At the level of abstraction that we have chosen to locate our analysis, "symbols" are not required "to be about" anything in particular, they are just the kind of units that a system of computation is capable of dealing with. As a matter of fact, we would argue that "aboutness"—understood here in a rather broad sense: To represent a concept, a percept, a motor instruction, and so on—is a derivative property of symbols, obtained from the connections they establish with one or another performance system—intentional/conceptual, sensorimotor, or whatever, as in Chomsky (1995) and subsequent works. A crucial implication of the previous statements is that the system of computation we have in mind is a general-purpose device, a representation-generating machine capable of acting on symbols from different external systems. From now on we will refer to this particular system, the core component of the structure that we are naming the Central Computational Complex (CCC), using the term "Computational System"—or, for the sake of simplicity, CS.

The next two crucial claims follow naturally from what has already been said:

---

[1] These are the premises of the representational/computational approach of cognitive science, as presented, among other places, in Putnam (1975 [1960]), Fodor (1975, 2008), Pylyshyn (1980, 1984), or Gallistel and King (2009).

1. Organisms differ in the connections that CS establishes with other, more specialized or domain-specific systems; and
2. CSs can be homologized abstracting away from these connections.

Summing up: CS is (i) a particular and autonomous system of the mind/brain of certain organisms—vertebrates, as a minimum; (ii) it is connected to different external systems—in this respect, the natural expectation is a high degree of species-specificity of ensuing CCCs; and (iii) it lacks an associated collection of symbols, and provides instead the elementary tools for computing behaviorally useful representations. This latter property amounts to saying that CSs, as such, are not subjected to the restrictions imposed by the content (semantics) and basic inferential relations (syntax) of symbols, quite the opposite of what happens in other domain-dedicated mental modules; on the contrary, they (CSs) impose on these modules a basic equipment of computational means, whose power constrains the richness and flexibility of the representations underlying behaviors in the relevant domains of experience. We are thus claiming that CSs belong to a level in the analysis of cognition corresponding to a set of basic mental operations directly provided by the biological substratum of the mind—or, in Pylyshyn's (1984) words, to its "functional architecture."

A note of clarification is in order: Given our characterization of CS, it is clear that it is not the same object that Chomsky has in mind when he thinks of a system of computation as the central component of FL (Chomsky 1995, and subsequent works). This is especially clear in Hauser et al.'s (2002) formulation of the same idea, where they identify this system with the Faculty of Language in a Narrow Sense (FLN), which they regard as a human and language-specific aspect of cognition. Note that, for Hauser et al. (2002), what is debatable is the human and language specificity of "recursion," a formal property of FLN. It is vis-à-vis this property, and not FLN, that they do not exclude the possibility that it could be an evolutionary co-option from some other faculty (navigation, numerical intelligence, social cognition, etc.) and that it could be present in the minds of other species. On the contrary, our own contention is that CS is a general purpose and extremely extended aspect of cognition. If pressed to maintain Hauser, Chomsky, and Fitch's terminological distinctions, we would then adopt the position of attributing to FLN a very broad range of species, subserving in each case a different array of abilities. We elaborate on this idea in Chapter 7.

Our expectation is thus that very disparate abilities are based on the workings of homologous CSs—for example, birdsong and human language, as certain molecular evidence, to be commented below, seems to suggest. In the case of human language, but not in the case of birdsong, the system of computation interfaces with an external conceptual system, but this is an irrelevant fact when it comes to establishing the natural identity of the corresponding CSs—see below for some further qualifications

Computational Homology 93

of this idea. It must be stressed, then, that we are not suggesting that birdsong and language are homologous capacities: We are just claiming that they are ultimately based on the same natural kind of structure/activity.

Let us see now how we conceive of the nature and functioning of this system in the case of a relatively simple form of birdsong, to wit the song of Bengalese finches (*Lonchura striata domestica*), of which we have a detailed formal analysis by Okanoya (2002)—see also Berwick et al. (2011). The songs of this bird basically consist of a series of "notes," represented by lower case letters in figure 5.1, subject to the following constraints:

1. Some notes can only be followed by some other specific note—for example, *b* must be followed by *c* and *c* must be followed by *d*, which means that these notes form fixed motifs within the song template;
2. Some other notes, however, can be followed, in different songs, by one or another different note—for example, *d* can be followed by either *b*, *i*, or *e*, but after *e* or the motif *im*, however, only *f* can appear, which means that at points like *d* the song is open to probabilistic transitions;
3. Finally, the motif *bcd*, following either *p* or *q*—ϵ represents the omission of a certain note or motif—allows the introduction of a new sequence obeying the constraints of the pattern so far described—a form of trivial "tail recursion"; see Fitch (2010: 78).

We thus conclude that, even if probabilistically at certain points, the possible transitions within the sequence of a particular song are always predictable: Each note always follows one or another note (three at the most). In addition, songs are also highly predictable in that recurrent sequences only obtain by adding a complete new (sub)sequence at the end of another one.

In order to understand our point it is crucial to tell two different aspects of figure 5.1 apart:

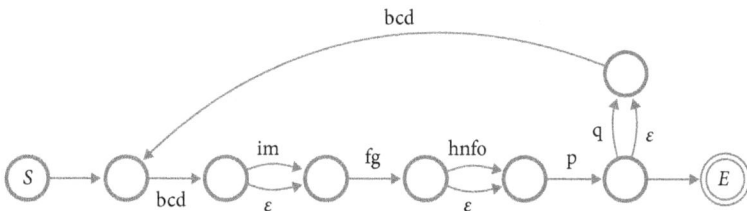

FIGURE 5.1 "Worms vs. Birds"
Possible transitions between the notes (lower case letters) that form the songs of Bengalese finches by means of a nondeterministic finite-state automaton. The introduction of each new note is only constrained by the previous one.
Figure based on Okanoya (2002). Caption: song title from the album *Sad Sappy Sucker*, by Modest Mouse (2001).

(i) The specific sequences that running this abstract transition system can lead to, which correspond to the (potentially infinite) set of songs that the bird is capable of performing—say, *bcbefgp*, *bcdimfghnfop*, *bcdefgpbcdbcdefgp*, and so on; and

(ii) The nature—or type—of the transitions that lead to this particular kind of sequences. Note that the next step in the process (change of state) is always determined by the sequence of notes generated in the immediately preceding transition and never by those sequences generated in transitions two, three, or more steps before the current one. Moreover, the set of states is always finite.

This distinction settled, the important thing to be aware of is that (i) is not an inherent part of CS, just (ii) is. This amounts to saying that the musical properties of the phrases that the bird is capable of composing—perceptible at the level corresponding to (i)—are foreign to this particular cognitive structure. Actually, (i) reflects the syntactic constraints governing the symbols of a particular domain of application of the system—in this case, musicality, or whatever we decide to name it. What we are contending is that, in the case of this organism, this particular domain is subserved by a system—CS—whose generative capacities are restricted to highly predictable forms of transitions between representational states—so (ii) reflects properties of CS, independently of any domain it applies to. From the point of view of this system—CS, music is just symbol manipulation and generation of representations—i.e., music without musical qualities.

Generalizing from the above observations, our main claims can be expressed as follows: (i) Language, just like birdsong, is subserved by a system—CS—for which language is devoid of grammatical qualities; and (ii) The systems in question are comprised within a larger natural class of such systems—i.e., of true homologues. It is true that birdsong and language show rather different properties even at this raw level of analysis—see next section, but regarding this question we further contend: (iii) That diversity at the computational level is of the sort expected among true homologues—remember, "The same organ under every variety of form and function." We turn to this question in the following section.

5.2 Issues of Complexity: Chomsky's Hierarchy and Natural Subclasses of Computation

Exhaustive and detailed characterizations of all the defining properties of specific computational phenotypes are impossible at a morphological level. However, we have at our disposal a powerful tool that makes it possible to identify the main features of systems of computation at an abstract level, and to elaborate concrete proposals about what structures and neural organizations could be associated with these features. In concrete terms, we will assume that there exists a minimum of four

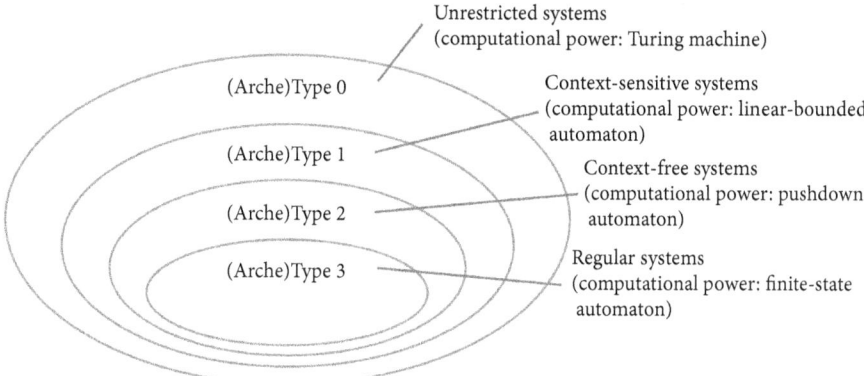

FIGURE 5.2 "Noam Chomsky Spring Break 2002"
Archetypal systems of computation are hierarchically ordered in terms of structural complexity: Highly ranked systems introduce symbol manipulation capabilities that are outside the scope of simpler systems.
Caption: song title from the album *The Cold Nose*, by Department of Eagles (2005).

basic types of CS—or "archetypes," following the conclusions of the previous section, in direct correspondence with the four levels of computational complexity of the Chomsky Hierarchy—figure 5.2.[2]

Note then that, despite the fact of not having a precise characterization of computational phenotypes at the morphological level (but see the next section for some proposals in this direction), we do have a precise abstract characterization of our computational archetypes. In fact, whatever the specific morphological properties of a phenotype, we know that, if it is associated with Type 3, its computational power will be equivalent to a finite-state automaton; if it is associated with Type 2, it will be equivalent to a pushdown automaton; if it is associated with Type 1, it will be equivalent to a linear-bounded automaton; and, finally, if it is associated with Type 0, it will be equivalent to a Turing machine, although, for reasons to be made explicit in Chapter 6, this type will be left out of our discussion.[3]

---

[2] We speak of a minimum of four archetypes, because we will for the moment stick to the original version of the Hierarchy (Chomsky 1959b). We are perfectly aware, however, that at a later date new levels were added to it. For example, Chomsky himself argued soon after for the necessity of distinguishing between strict context-sensitive systems and systems capable of generating any recursive system (Chomsky 1963); Aho (1968) described indexed systems within the complexity space originally reserved to Type 1 systems—see also Hopcroft and Ullman (1979: Ch. 14), whereas Aravind Joshi and collaborators—see Joshi (1985), Joshi et al. (1991), and Vijay-Shanker and Weir (1994)—added new compartments to this space with the extended context-free systems or, alternatively, mildly context-sensitive systems. What follows, however, does not particularly hinge on the exact number of types—although we will come back to these new developments—and for ease of exposition we will stick to a space divided into four basic compartments—with FL somewhere within the "lower" area of Type 1 systems.

[3] Note that our archetypes define a collection of equivalence classes to which natural computational systems can be assigned given their computational power and independently of fine points of detail

The simplest computational devices are Type 3 or regular systems, whose equivalent in the theory of abstract machines is the finite-state automaton. These systems have the power of generating sets of structurally very simple strings—or "languages."[4] Thus, for example, the following languages are regular in the intended sense:

1. a^*—any sequence of zero or more as;
2. a^*b^*—a possibly null sequence of as followed by a possibly null sequence of bs;
3. $a^n b^m$—a non-null sequence of as followed by a non-null sequence of bs; and
4. $\{a, b\}^*$—that is, the one constituted by sequences of as and bs of any length and in any order.

Note that the sets defining each of these languages only incorporate constraints over any two contiguous symbols in a string: in (1), an a can only be followed by another a; in (2), an a can be followed either by another a, or by a b—but at just one point of the string, and a b can only be followed by another b; in (3), an a must be followed either by another a, or by a b—but at just one point of the string, and a b can only be followed by another b; and in (4), an a can be followed either by another a or by a b and a b can be followed either by an a or by another b—at any point of the string. Note also that these languages differ in properties that have to do with probabilistic transitions: In (1) the string can end at any point "or" add another a—another instance of trivial tail recursion; in (2) and (3) there is additionally a point at which both an a "or" a b can be the following symbol; and in (4) as "or" bs can freely alternate at any point of the string. Notwithstanding, they belong to the same area of the hierarchy, (Arche) Type 3, in that even when sequences can intuitively be said to contain subsequences of different symbols—as in *aaaabb*, belonging to languages (2), (3), and (4), or similar strings—no relation actually holds between or across the relevant subsequences. A quick glance at these languages is sufficient to see that linear order is not a problem for this kind of system, as it is perfectly possible to build grammars—or automata—capable of generating sequences where symbols follow a strict order.

There is a sufficient guarantee of adequacy for regular grammars to capture the complexity of birdsong strings like that of Bengalese finches—figure 5.1 above, but not of utterances of natural languages. In fact, Chomsky, in a brief note (Chomsky 1956b), already demonstrated that a language like $a^n b^n$—a sequence of as followed by a sequence with the *same* number of bs—is beyond the generative power of a finite-

---

concerning their actual architecture. In this chapter and the following two we will be dealing with matters of computational complexity in a rather informal way; for a more comprehensive and formal presentation, the reader is referred to the Appendix at the end of the book.

[4] It is important not to confuse the term "language", as we use it here in the context of formal language and automata theory, with its traditional meaning in the field of generative linguistics; see Appendix for details.

state system. The datum to which we need to pay attention here is that the complexity of this new language has nothing to do with the relative order of both substrings, but rather with the fact that both substrings must be of the same length, which is equivalent to saying that there exists a dependency relation between both substructures. In other words, in order to be sure that both substrings will be of the same length, we need some device to keep track of the number of symbols used during the process of construction of the first substring, so that we can access this information while we are building the second one. In a nutshell, we need *memory*, a resource that is not available in a finite-state automaton. As Chomsky points out in the reference cited above, natural languages have plenty of this kind of dependency relations, which automatically invalidates the ability of finite-state systems to capture some of the most basic properties of human language. For example, constructions of the *if... then*-type, relative clauses, or sentential complementation, are just three of them:

(1) a If$_i$ the boy likes red apples then$_i$ I will give some to him
    b The boy$_i$ that likes red apples is$_i$ here
    c The boy said$_i$ that he likes red apples with$_i$ a clear and loud voice

Note, by the way, that what is important in these cases is not that both strings are of the same length, but that some dependency relation holds between two elements separated by an arbitrary long sequence of symbols:

(2) a If$_i$ the boy with a blue hat likes red apples then$_i$ I will give some to him
    b The boy$_i$ that likes red apples and sweet oranges is$_i$ here
    c The boy said$_i$ that he likes red apples after a good lunch with$_i$ a clear and loud voice

As readers have probably noticed, this kind of dependency is a key property of linguistic patterns of organization, beginning at the level that we referred to as the "Tesnièrean layer" in section 1.1.2.2. As we will see presently, the number and nature of these dependencies are critical factors at the time of assessing the degree of complexity of a language.

In the light of these results—well known, as we have seen, since the 1950s—research in the field of the formal complexity of natural language turned its attention to Type 2 and Type 1 systems. Let's go back, then, to the language $a^n b^n$, which, as we saw, is not a Type 3 language, but a Type 2 language. Depending on the kind of structural description we want for, say, the string *aaabbb*, a context-free grammar offers us a number of alternatives, of which we only contemplate the following:

(3) [a$_i$ [a$_j$ [a$_k$ b$_k$] b$_j$] b$_i$]

Here, sub-indices indicate the presence of some dependency between the elements sharing the same sub-index. As we will see presently, the source of complexity is not in the number of individual dependencies between pairs of symbols, but in the global pattern of relations among them. Note that in (3) dependencies are strictly nested, and a Type 2 system is perfectly capable of dealing with constructions with multiple nested dependencies—or, for that matter, with sets of independent nested dependencies, that is, $a^n b^n c^m d^m$, $n \neq m$, is also a Type 2 language. Dependencies of this type are actually the point that sentences in (1) and (2) above are designed to illustrate, as represented in the schemas in (4):

(4) a [[$_{if}$...] [$_{then}$...]]
    b [[...[$_{that/rel}$...]]...]
    c [...[$_{that/comp}$...]...]

As we pointed out earlier, the key for dealing with these patterns of organization is in memory, a resource that is available to a pushdown automaton, but not to a finite-state machine. The pushdown stack in a pushdown automaton supplies the additional workspace where we can store those symbols we have generated—for example, three *a*s—and which we pop out as we add *b*s to the string: For each *b* we add to the string, we pop an *a* out of the stack, such that, when the stack is empty, the process is over. Given the structure of the stack, which follows a "first-in/last-out" regime, we can see that nested dependencies fall within the power of Type 2 grammars since, when we write the first *b*, we pop out the last *a* that went into the stack, and so on, until the point at which we write the last *b* and pop out the first *a* we stored in memory—this guarantees that dependencies never cross, so the system is simply required to deal with one such dependency at a time. Obviously enough, for processing strings obeying the patterns in (4) a more sophisticated regime is required, but the point of taking (3) as a reference is to demonstrate that such a regime falls, at a minimum, within the area of Type 2 grammars.

Suppose now that dependencies are organized as in (5):

(5) $a_i a_j a_k b_i b_j b_k$

Note that in this case the dependencies are crossed—another pervasive feature of natural languages, such that the first *a* is related to the first *b*, the second *a* with the second *b*, and so on. These dependencies are found in Dutch and in some varieties of Southern German, where the English construction ... *that John wants to let Mary read the book* may be expressed, for example in Dutch as ... *dat Jan Marie het boek wil laten lezen*—lit. "that John Mary the book wants to let to read;" figure 5.3.

This kind of structure is beyond the processing power of a pushdown automaton, as are more complex languages like $a^n b^n c^n$. Without for the moment going into great detail—but see Weir (1994) and Joshi and Schabes (1997: Sect. 7), what we need here is a more powerful automaton, one we can get just by improving the capabilities of the memory system, extending and restructuring it in such a way that it will be able to

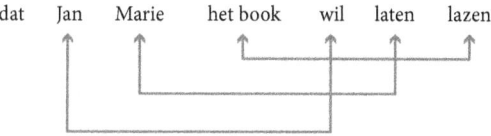

FIGURE 5.3 "Amsterdam"
A fragment of Dutch, exhibiting cross-serial long-distance dependencies.
Caption: song title from the album *A Rush of Blood to the Head*, by Coldplay (2002).

create additional stacks to store data any time this is required by the computation.[5] Note that cross-serial dependencies are a pervasive in linguistic expression, as it is easy to grasp by pinpointing some of them in the sentences in (1)—they are added to the long-distance dependencies that obtain by nested embedding:

(6) a If$_i$ the boy$_j$ likes red apples$_k$ then$_i$ I will give some$_k$ to him$_j$
    b The boy$_{i+j}$ that likes$_j$ red apples is$_i$ here
    c The boy$_j$ said$_i$ that he$_j$ likes red apples with$_i$ a clear and loud voice

Complex matrices of cross-serial long-distance relations of this sort are the key property of the level of linguistic organization that we referred to as the "Chomskyan layer," which is now easy to see why it necessarily requires a memory regime within the area of Type 1 grammars.

Thus, and on the basis of the preceding discussion, it is clear that the progression up the scale of complexity is a function of the changes introduced in the memory system, *with no other modification of any fundamental property of the computational system being necessary*. This observation puts us in a position not only of being able to subcategorize CSs according to the computational archetypes corresponding to the spaces in figure 5.2 above, but also of being able to determine the kinds of alterations

---

[5] This is an extremely intuitive characterization of the embedded pushdown automaton, which is equivalent to a mildly context-sensitive system. We are using this example here, instead of that of the linear-bounded automaton, because as Weir (1992, 1994) has shown building on earlier work by Khabbaz (1974), pushdown automata constitute a general model of automata, of which the classical pushdown automaton—with one stack—and the embedded pushdown automaton are only two particular cases, the simplest ones within a scale of increasing complexity definable exclusively in terms of improvements introduced in the storage system. Weir's results in fact go well beyond that, since they can be generalized to the whole family of languages made up by Type 2 and Type 1 languages in the Chomsky Hierarchy and which constitute a natural class within the Hierarchy, sharing a number of interesting computational properties—e.g., recognition in polynomial time and decidability, among others. Weir's work defines a sub-hierarchy within the old hierarchy, in which we observe a progression towards higher degrees of complexity definable just in terms of the levels of embedding of the memory stacks—i.e., stacks of stacks, stacks of stacks of stacks, and so on, a particularly relevant mathematical result for the proposals we develop in this book.

necessary to "jump" from one space to the other, a question we will deal with at length in Chapter 6.

5.3 From Architecture to Anatomy (and from Anatomy to Molecules)

In previous sections we have argued that homologies can be established in terms of activity—for example, computational activity, as organic activity in the intended sense, directly follows from structural organization. Therefore the next question is: Of which physical structure is "computation" the proper organic activity? As announced at the very beginning of the previous section, an understanding "in the abstract" of the powers of a particular CS can help us to elaborate concrete proposals about what structures and neural organizations can be hypothesized as its physical basis. In general, we saw that CSs are "sequencing machines;" more specifically, we also saw that, starting with Type 2, these machines are associated with a "working memory" space. These two components comprise the basic "functional architecture" of a computing device. As for the physical implementation of such a device, we adhere in this section—with some minor adjustments—to the model recently proposed by Lieberman (2006) under the name of "Basal Ganglia Grammar," in which the following correspondences between functional components and anatomical sites are defended:

1. A *pattern generator* (or *sequencer*)—whose inhibition/excitation mechanism is located in the basal ganglia; and
2. A *working memory space*—located in Broca's area (Lieberman 2006: 207–209).

These are, respectively, the subcortical (1) and cortical (2) components of a circuit that Lieberman defines in functional terms as an *iterative sequencing machine*, which is at work when we walk, talk, or understand a sentence (Lieberman 2006). Assuming this global picture, we wish, however, to introduce a couple of qualifications over the adoption of this neuroanatomical structure as the basis of our model for the computational system underlying language.

First, the basal ganglia comprise a complex anatomical structure that appears to participate in several cortico-subcortico-cortical circuits associated with the regulation of different aspects of mobility, cognition, and emotion (figure 5.4, left). Following Cummings (1993), Lieberman (2006: 163–167) estimates that the so-called "dorsolateral prefrontal circuit" is the one involved in the programming of the motor control of speech, in sentence comprehension, and in other aspects of cognition. This circuit projects from the prefrontal cortical area towards the dorso-lateral area of the caudate nucleus, the lateral dorso-medial area of the globus pallidus, and the thalamus which, in turn, projects back to the prefrontal cortex (figure 5.4, right). We will hypothesize—following Lieberman—that this is in effect the circuit that language uses as a computational system.

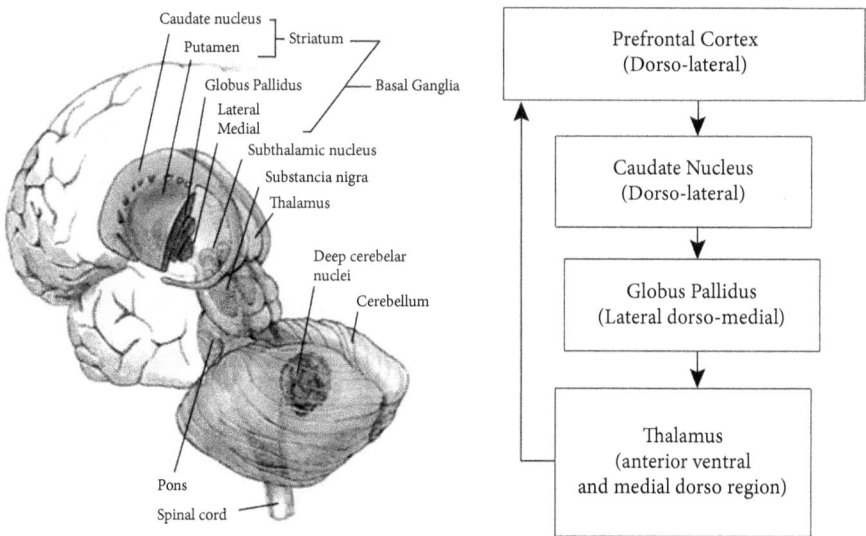

FIGURE 5.4 "Something Should Be Said"
The dorso-lateral prefrontal circuit, linking the prefrontal cortex with some components of the basal ganglia, is involved—although not exclusively—in the motor programming of speech and sentence comprehension. Other similar, and partially overlapping, cortico-subcortico-cortical circuits participate in other aspects of cognition and behavior.
Caption: song title from the album *Monster Head Room*, by The Ganglians (2009).

According to Lieberman, the functioning of this circuit proves crucial in very disparate tasks, some of which imply the execution of complex motor patterns—as in speech or music execution, while others are internally performed—such as sentence planning and comprehension, music perception, or certain classic examples of problem solving, for example the Tower of London test. The common feature of such a heterogeneous collection of activities seems to be their underlying sequential organization, which is a most natural reason for the recruitment of a structure consisting—as its basic component—of a pattern generator. Strong pieces of evidence in favor of the idea are mainly taken by Lieberman from the study of linguistic and cognitive deficits typically associated to Parkinson's disease, a degenerative syndrome affecting the basal ganglia, where patterns of the above-mentioned types become more or less subtly, but progressively disintegrated. Lieberman also notes that similar effects are typically observed in cases of Broca's aphasia, which—according to him— "does not occur, absent subcortical brain damage" (Lieberman 2006: 176). Observations similar to these extend, for example, to the case of music appreciation, where the role of the basal ganglia in rhythm perception has been firmly established through the study of patients with Parkinson's disease (Grahn 2009).

Now, and this constitutes our second qualification of the Basal Ganglia Grammar model, we contend that limiting to Broca's area the localization of the working memory space is an oversimplifying assumption. The involvement of Broca's area in the system of computations of the FL, or CCC_{HUMAN}, in our favored terminology, is unquestionable (Embick et al. 2000; Grodzinsky 2000; Moro et al. 2001; Musso et al. 2003, among others). However, it is quite plausible that this structure is part of a larger cortical circuit with bidirectional projections between the frontal and parieto-temporal areas, within a system of working memory networks such as the one postulated by Aboitiz et al. (2006).[6] The matter is not entirely clear. However, for the purposes of this book, it is important for us to localize the working memory of the computational system within the cortical component, more or less large, of the anatomical structure making up such a system.[7]

In a way, using the label "Basal Ganglia Grammar" to name this system could somehow seem misleading, given that FL owes its computational powers to the cortical component—i.e., to memory, and not to the subcortical constituent. However, seeing things that way would represent an anthropocentric view on the matter, because the basal ganglia are the universal component of the system and, for this reason, the element that gives us the opportunity to homologize systems of computation irrespective of the particular level of complexity they attain. Thus Lieberman's terminological decision happens to be felicitous. This is particularly so when confronting the fact that variants of the same transcription factor—FOXP2/FoxP2—seem to be at work in the development of the human and the avian versions of this system. In the case of birds, expression of the gene coding for this factor is registered in the so-called area X—a component of the anterior brain circuit associated with the learning of birdsong, which is part of the avian structure that is homologous with basal ganglia in humans. Significantly enough, this gene's activity also extends to the frontal cortex in the case of humans. Spontaneous or induced—in the case of birds—point mutations in these genes are linked to impairments affecting singing and verbal abilities, respectively.[8] These facts give rise to a picture fully consistent with the

---

[6] See also Aboitiz and García (1997). On the role of the frontal lobes in language processing, see the review by Friederici (2002), as well as the studies by Shtyrov et al. (2003) and Pulvermüller and Assadollahi (2007).

[7] For a partially divergent model, see Ullman (2004), where the assumption is made that the computational aspect of language utilizes a system of "procedural memory" distributed across the frontal and parietal cortical areas, the basal ganglia, and the cerebellum. In Ullman's model, however, no computational specializations between a sequencer and a working memory are assumed, and the basal ganglia are seen as responsible for the stimulation and inhibition of the memorized patterns across the whole brain circuit.

[8] See Ferland et al. (2003), Haesler et al. (2004), Haesler et al. (2007), Lai et al. (2001), Lai et al. (2003), Liégeois et al. (2003), Miller et al. (2008), Rochefort et al. (2007), Takahashi et al. (2003), Teramitsu et al. (2004), and Teramitsu and White (2006). See also Benítez-Burraco (2009) for a comprehensive and up-to-date overview of data concerning FOXP2, and Piattelli-Palmarini and Uriagereka (2011) for some speculations as to how to interpret all these data.

proposed computational homology, both regarding its invariant and variant aspects. For the time being, let this brief observation suffice as a piece of empirical support for our general framework. Anatomical and molecular aspects of the proposal will be more nuanced and touched on in Chapter 6.

5.4 Computational Phenotypes and Multilayered Homology

The idea of "hierarchy" captures a basic and extremely pervasive form of organization, most probably extending to every domain of experience. From the inorganic to the organic, from the natural to the institutional, from Cosmos to the ultimate components of matter, there seems to be no escape from the idea that a universal pattern of relative inclusiveness is present domain after domain (Simon 1969). It is generally assumed that such a pattern is actually a basic principle of organization in the biological realm (Hall 1994; Newman 2003), where several hierarchically organized levels of structure—from molecules to observable behavior, through cells, tissues, organs, organic systems, and physiological activity, probably a non-exhaustive list—are routinely taken for granted in descriptions and explanations. The main claim of this chapter is that there also exists a "computational level" of description/explanation, corresponding to the activity—in the sense of Love (2007); see also Striedter and Northcutt (1991: 182, fn. 3)—of structures of the nervous system, and thus comparable to the physiological level in which analysis of other organic structures is customarily and non-problematically located.

A different and non-trivial question is, however, that of establishing what the causal link is between the different levels at which we can place the description of some particular natural structure, be it in the same or in different organisms. This problem leads directly to another important conceptual issue, concerning the relation between the notions of "hierarchy" and "homology" (Roth 1991; Hall 1994; Minelli 1998, 2003; Wake 2003). Striedter and Northcutt (1991), for example, defend a relation of "quasi-independence" between different levels of biological analysis and, consequently, between the homological relations settled at each level. Levels—in their case, genes > developmental processes > morphological structures > physiological functions > behavior—are obviously not independent in that they are causally linked; however, they are independent in that units at different levels are not isomorphic—see Hall (1999: Ch. 21) for a similar tenet. As a consequence, homologies can be established at a particular level without requiring the existence of equivalent homologies at lower and upper levels of analysis. Such a degree of inter-level autonomy—with each level consisting of units extraneous to neighboring levels—gives rise to such a massive "chain of superveniences"—and a "pluralism of properties" of sorts—that it seems difficult to integrate it into an acceptable materialist view of nature.[9]

As an alternative, we suggest that levels of analysis are isomorphic, so special homologies established at some level of analysis l^n are also verified at levels l^{n-1} and l^{n+1}. With this suggestion, levels lose autonomy, but the unity of a given organic structure is reinforced, because it does not atomize into disparate entities with a change of perspective in the analyses.[10] We think that this point of view, even if "ontologically humble"—or maybe because of its ontological modesty, is more in agreement with the goals of biological theory. In any event, the main advantage of this position has to do with the foundations of a solid ahistorical concept of homology. "Homology"— according to Arendt (2005: 186)—"is about history and cannot be proven." However, once a "principle of isomorphic correspondence" between levels of analysis of a given biological structure is raised, this sort of skepticism around homologies can be overcome, because they don't need to be proved by means of historical findings, but by means of convergent results at different levels of analysis. Homologies have of course history—that of the structures under consideration, but they are not about history—they are just about structures—and, thus, can be proven. The above-mentioned principle of correspondence is a strategy with which the concept of homology is clearly reinforced.

For our own concerns, it is enough to consider a hierarchy of just four levels of analysis: developmental system > morphological structure > physiology > computational activity. Developmental systems incorporate, but are not exhausted by genes—see the next chapter for details. We also understand that developmental processes are indistinguishable from the factors composing the systems underlying them—they correspond, respectively, to the dynamic and static aspect of a single level of analysis. Physiology is also a relevant level, intermediary between morphology and computation, and often the most direct cause of many complex behaviors; we will not enter into a detailed analysis of this level as it is not crucial for our arguments here, but see Ketterson and Nolan (1999), Costa and Sinervo (2004), and Garland and Kelly (2006) for some comprehensive overviews. For reasons by now clear, we exclude behavior as a true level of evolutionary analysis. Or, to be more precise—and following Klopfer's (1973b: 60) take on the issue: "Behavior is not a thing, defined and determined by a discrete locus on a DNA molecule. It is a *process* that derives from a series of interactions, which at times can achieve a certain level of predictability and stereotypy." To be sure, it is important not confuse the fact that behavior is

---

[9] We consider this form of pluralism problematic in that it multiplies the difficulties of, for example, "dualism of properties" in the analysis of mental phenomena, arguments, and actual support of the idea— as in Chalmers (1996)—notwithstanding.

[10] This position is in agreement with Chomsky's (1980) idea that levels of analysis—even if associated with different types of vocabulary (physical or abstract), are just modifications of our point of observation. See also Balari and Lorenzo (2009: 19–21).

one of the highest levels of biological organization and hence one of the most visible ones to selection (Garland and Kelly 2006) with its putative causal powers, which are in all likelihood very limited or altogether absent. Behavior is caused, but it is hardly ever itself a cause.[11] We thus contend that anything interesting that can be said about behavior is actually covered by whatever can be said concerning the four levels listed above, given the dynamic perspective advocated here. Note, moreover, that a strong methodological principle for the analysis of behavior is emerging, in perfect consonance with our discussion in Chapters 3 and 4. We will fully illustrate this principle in Chapter 7, but let us point out here that what is relevant in this case are not the functions and putative uses of some behavior but, rather, the complexity of the underlying patterns that can be inferred from its observation, which provide us with important information concerning the cognitive resources that are put to use in order to bring about some behavioral act, like, for example a birdsong or a sentence in a natural language. It is in this sense that behavior is irrelevant or, rather, that it is relevant only to the extent that it is a window through which we have access to the computational activities that underlie it. With all these pieces put together, we can advance our proposal.

We would like to suggest that figure 5.2 depicts the archetypes from which we can generalize homologies as regards the computational capabilities of different species. These archetypes relate to an abstract "draft organ"—in the sense of Minelli (2003)—with a basic structure "pattern generator + working memory," with the latter component—working memory—as the *locus* of diversification along the scale "finite state," "pushdown," "linear-bounded"—see Chapter 6 for details—and with the exclusion of "Turing machine" from this scale of measure. We suggest "computype" as a short term to refer to these variants. From a certain point of view there exists just one computational archetype—corresponding to the "draft organ" of computation; from an alternative point of view, however, there exist three alternative computypes: CompuT3, CompuT2, and CompuT1. There is no contradiction in this—Nature is hierarchically organized.

If we move to the next lower level of analysis, each computype corresponds to a particular "phylotype". We use the term here in a sense only slightly modified from that of, for example, Minelli and Schram (1994)—see also Minelli (1996), based on the concept of "phylotypic stage" of Slack et al. (1993). This family of concepts refers to a particular stage of embryonic development—thus, to a morphological structure, with certain well-defined locations and positions, albeit still subject to processes of differentiation, like the particular structures or appendages to be developed there. Again, the general concept of "phylotypic stage" comprises a collection of alternative

---

[11] Or in any case the initiator of a causal chain. Behavior may contribute to the fixation of, say, some physiological novelty, but the novelty had to be there before to make the behavior that would eventually fix it possible.

phylotypes—"platytype," "cyclotype," "malacotype," "trimerotype", or "arthrotype;"[12] see Minelli and Schram (1994). As for our own proposal, we claim that the general or common computype—i.e., "pattern generator + working memory"—relates to a general or common phylotype as well, a morphological structure comprising the basal ganglia and a particularly atrophied component of the pallium, with which the basal structure is to develop most of its nervous connections. The exact identity of that component is open to a certain range of inter-specific variation: In the case of mammals it is the neo-cortex; in the case of birds, the dorsal ventricular ridge.[13] Note that, at this level, we are still facing "draft" morphologies, whose developmental fate is still open to a considerable degree. Besides, as observed by Striedter (2005), at this level connections are the best clue to be suspicious about the existence of homologies—note that our principle of isomorphic correspondence applies to levels, not to species, and that inter-specific diversity is part of the definition of homology. According now to a "principle of maximal congruence"—like the one suggested by Young (1993), the common archetype at this level would thus consist of (i) basal ganglia, (ii) an atrophied component of the pallium and, crucially, (iii) a pathway of reentering connections between (i) and (ii). As in the upper level of analysis, three alternative phylotypes—PhyloT3, PhyloT2, and PhyloT1—can also be hypothesized at this level, having to do with the power of the physical materialization of the memory component (ii)—or maybe the connective capacities of the nervous pathway (iii). With each particular phylotype a qualitatively different form of computation—i.e., a computype—is obtained.

Isomorphic inter-level correspondence implies that homologies hypothesized at higher levels of analysis must ultimately obtain confirmation at the lowest developmental level. A shared collection of factors and constraints on development is a *sine qua non* condition in order to deem homologous a certain group of structures/activities (Roth 1984, 1988; Wagner 1989a, 1989b). We would like to express this requirement by saying that each phylo/computype must correspond with a particular "zootype," extending to the realm of computation—Slack et al.'s (1993) concept—see also Schierwater and DeSalle (2001).[14] A particularly challenging circumstance in this

---

[12] With the necessary adjustments, the "vertebral archetype", as envisioned by Owen (1848), could be added to this list.

[13] The dorsal ventricular ridge is anatomically homologous to the amygdala and claustrum of the mammal pallium—see Striedter et al. (1998), Striedter (2005), and Avian Brain Nomenclature Consortium (2005).

[14] We consider our own concepts of "zootype" and "phylotype" to be extensions of their homologues in Slack et al. (1993), Minelli and Schram (1994), or Minelli (1996) for two different reasons. On the one hand, the concepts refer in all these works to early stages of embryonic development with a high impact in determining the basic body plan of different groups of animals. In our case, they refer to relatively late and more localized episodes; i.e., more restricted specifications of the general body plan. On the other hand, according to Deutsch and Le Guyader's (1998) hypothesis, the primitive and most basic function of the zootype—even more basic than segmentation—is to establish an elementary set of neural connections

case is the above-mentioned fact that different anatomical structures are most probably recruited in different developmental scenarios—as in birds and mammals. In any event, some pieces of evidence point to the idea of a common underlying zootype as a reasonable conclusion, such as the existence of (i) homologous cell populations in the intra-cephalic circuitries of these species irrespective of the diversification of the corresponding structures—see Karten and Shimizu (1989); and of (ii) homologous transcription factors (FOXP2/FoxP2) whose regulatory activity extends to parts of the structure in both cases.[15] As in the case of higher levels of analysis, different alternative zootypes can also be envisioned at this one—ZooT3, ZooT2, ZooT1, corresponding to modifications in development significantly increasing the memory resources underlying qualitatively different systems of computation.

5.5 Conclusions

Evolutionary reconstructions crucially depend on the concept of "homology." In this sense, the case of the evolutionary study of language seems somehow exceptional, because it is customarily assumed that FL has no true homologues (Chomsky 1968; Bickerton 1990), an idea that we reject. In the first section of this chapter we argued that this (false) problem of continuity is mostly due to the fact that, in the case of language, the comparative method has been applied to observable behaviors (namely, the so-called "communicative behaviors"), a level of analysis whose units do not constitute true natural kinds and at which no true homological relations can be settled. In this chapter we have further argued that by applying the comparative method to the core component of FL—CCC_{HUMAN}, and under a particular interpretation of its computational system, homologies proliferate and, consequently, the evolutionary explanation of language becomes a task comparable to that of explaining any other biological character. Our approach relies on an innovative concept of "computational homology," which we have justified as a natural extension of the idea of "homology" established on structural or molecular grounds. The road is now open to a new and—we hope—more fruitful approach to the evolutionary origins of language.

---

along the body axis. In this sense, our own concept is literally a relatively late—and more specialized—extension of this elementary function.

[15] Although the developmental picture at the molecular level is almost certainly going to be much more complex, given the recently discovered key regulatory role of non-coding RNA, especially in brain development; see Mehler and Mattick (2007), Mattick (2007), Mattick et al. (2009), Gräff and Mansuy (2008), and Singer et al. (2010).

6

Introducing Computational Evo Devo

"Embryonic Journey", *Surrealistic Pillow*, Jefferson Airplane, 1967

So far we have reached a number of conclusions worth remembering before trying to frame them into an evolutionary model of explanation. We first defended a view on FL according to which it consists of a general-purpose Computational System (CS) plus a series of interfaces with different external or peripheral systems. We suggested using the term Central Computational Complex (CCC) to refer to this organic system as a whole. The motivation of this terminological twist was twofold: First, it is useful to underscore the idea that this very system underlies many other typically human activities—so "language" is just one of its observable outcomes; and second, it suggests that there exist non-human versions of this system—an idea that seems more difficult to digest preserving the anthropomorphic connotations of the FL concept. A logical consequence of this first move is that there exist two different loci at which interspecific variation affecting this structure may occur: (1) The relative power of the corresponding CS, measured according to standard criteria of computational complexity; and (2) the set of peripheral systems with which CS interfaces in each particular CCC. Comparative data regarding both points will be offered in Chapter 7. For the time being, we will concentrate on the issue of how a particular complexity level and a particular pattern of interface connections can be evolutionarily attained, a task that we consider feasible once we have identified the anatomical correlate of CS—a cortico-subcortico-cortical circuit comprising parts of the basal ganglia and an atrophied component of the pallium, and we have reasonably grounded the idea that the activity of variants of this organ patterns with the levels of complexity of the Chomsky Hierarchy—along the scale finite-state < context-free < context-sensitive. This latter contention is particularly important because it boils down to the claim that the space of computational variation is a discontinuous one, as each particular computational phenotype is not just a quantitative upgrading of less complex phenotypes. Rather, different phenotypes imply different types of computational activity, in a strictly qualitative sense. This is a very important conclusion of the previous chapter, because it establishes a severe constraint on models of evolution capable of explaining the diversification of CCCs. A complementary constraint to this one is that variation affecting the complexity of computations is irregularly distributed across taxa, a question that will be touched on in the next chapter. We advance it now as a further constraining condition to the model to be spelled out in the following pages.

Considering this conclusions and constraints, we begin to articulate in this chapter an internalist explanatory model for the evolutionary origins of the computational system subserving the human Central Computational Complex (CCC_{HUMAN}), inspired by ideas worked out during the last few decades in the field of Evolutionary Developmental Biology (Evo Devo). In particular, we will adhere to the general framework of "morphological evolution" elaborated by Pere Alberch (1954–1998) in a series of papers published in the eighties and nineties of the last century—Alberch (1980, 1982, 1989, and 1991)—by extending Alberch's original proposals to the realm of cognition.

Evo Devo aims at explaining the origins and evolution of natural designs (phylogeny) by means of hereditary perturbations affecting the development of organisms (ontogeny)—see Hall (1999, 2001), Raff (2000), Gilbert (2003a), Hall and Olson (2003a), Robert (2004), Carroll (2005), García-Azkonobieta (2005), Laubichler and Maienschein (2007), Müller (2008), Minelli (2007), and Arthur (2004, 2011). In a nutshell, the essence of Alberch's proposal against this general background can be summarized in the following statements:

1. The development of organisms is based on systems comprising different morphogenetic parameters—not just genetic factors;
2. Interactions among these parameters are complex and they correlate non-linearly with phenotypic outcomes—i.e., a particular parameter can be continuously changing without any observably significant consequence;
3. However, once a certain threshold—or critical value—is reached, a minor change in the parameter in question can be enough to introduce radical phenotypic manifestations;
4. These minor but far reaching perturbations on development—which do not necessarily imply new developmental factors or new patterns of interactions within the system—can eventually attain evolutionary significance; and
5. Morphogenetic parameters set the limits of the forms attainable from a particular developmental scheme. Thus, the direction that development/evolution

can take from a particular transitory phenotypic state is strongly constrained by the "geometry" of the parametric space so defined.[1]

The idea that evolution is strongly constrained by the very same factors that also strongly constrain the development of individuals is common ground for every Evo Devo-oriented approach on evolutionary phenomena. Conversely, this idea is not congenial with the more classical stances of neo-Darwinian selectionism in that, according to the adherents of the Modern Evolutionary Synthesis (Modern Synthesis), natural selection acts on the diversity randomly introduced into populations by point genetic mutations, totally unrestricted in scope—see the classical formulations of Morgan et al. (1915), Dobzhansky (1937), Mayr (1942), Huxley (1942), and Simpson (1944); see also Mayr (1963) and Mayr and Provine (1980) for general overviews.[2] Consequently, natural selection is believed to be the only creative force capable of organizing an otherwise amorphous material. Evo-Devoists, in contrast, attribute an important part of this creative capacity to constraints acting upon development and limiting the scope of attainable designs. Natural selection is consequently redefined as a mechanism just filtering out those designs that fit ecological and populational conditions more efficiently—see Goodwin (1994: 143) and Wagensberg (2004: 125), as well as Alberch (1980: 664), Oster and Alberch (1982: 455), Alberch (1989: 46–48), and Alberch (1991: 16).

In this chapter we contend that the Evo Devo theses thus far advanced can be readily extended to the evolutionary study of the nervous system and cognition—see Griffiths and Stotz (2000), Amundson (2006), Finlay (2007), and Griffiths (2007), for some programmatic attempts in this direction. Our main claim is that the different levels of computational complexity reflected in the Chomsky Hierarchy—figure 5.2—are the possible phenotypes of a cognitive parametric space defined by a restricted set of morphogenetic factors or parameters. These parameters are non-linearly related to the development of the neural resources that supply the memory requirements of each computational archetype—in the technical sense defined in the previous chapter. We argue moreover that the (mildly) context-sensitive power that can be assigned to the computational system of CCC_{HUMAN} is an emergent consequence of minor perturbations affecting the development of a cortico-striatal circuit, once a certain morphogenetic parameter reaches a critical value in the course of human evolution. Thus, the adoption of this computational regime by FL can be seen as an

---

[1] See Pigliucci (2010b) for a contemporary (positive) assessment of Alberch's ideas from the standpoint of Evo Devo. The notion of "parametric space" has its rough synonym in that of "phenotype landscape" defined in Rice (2008).
[2] The extent to which Evo Devo represents a challenge to neo-Darwinism is a hotly debated issue. Thus, Fodor and Piattelli-Palmarini (2010: Part I) present a rather gloomy perspective for the future of the Modern Synthesis; see Laublicher (2010) and Minelli (2010) for a balanced exposition of the different issues involved in the debate.

evolutionary outcome strictly channeled by the organic conditions settled by the system of parameters at hand.

We see our proposal as an alternative view to the one held by contemporary evolutionary psychology, whose practitioners defend the view that the mind is a collection of purpose-specific modules, each one an adaptively bespoke answer to environmental—i.e., external—conditions; see Pinker (1997), Plotkin (1997), or Buss (2007), as well as the applications of Pinker and Bloom (1990) or Jackendoff (2002) to the evolution of FL.[3] The internalist stance put forward in this chapter is on the contrary, based on the idea that the internal organization of the mind is in itself a constraining system that biases evolution in favor of certain forms of cognition and limits the power of the environment in the shaping of the organic design of minds.[4] This does not mean that environmental and populational factors can be put completely aside by internalist-oriented theories. On the contrary, no evolutionary theory ignoring the external factors acting as selective criteria for the diversity independently brought into being would be complete. In this sense, the ideas in this chapter complement the proposals sketched in Chapter 2, according to which the originally maladaptive character of this feature of the human cognitive phenotype would, somehow paradoxically, have played a crucial role in its fixation as a species feature. This is another aspect of our evolutionary recipe of CCC_{HUMAN} in which we strongly depart from externalist-oriented—or adaptationist—views on the evolution of human mind.

The first section of the present chapter is devoted to presenting the basic tenets of Evo Devo along with Alberch's model of morphological evolution. In the final section we work out our extension of this model to cognition and its application to the evolutionary origins of the human Computational System (CS_{HUMAN}).

6.1 Evolution through Development

> Development defines the realm of the possible.
>
> Pere Alberch, 1980

A somewhat unsuspected episode in the recent history of biology was the divorce between the study of individual development—the classical subject matter of embryology—and that of evolution at the species level—in charge of population genetics for the most part of the twentieth century (Morgan 1932; Dobzhansky 1937; Hamburger

---

[3] Evolutionary psychologists tend to show a rather belligerent and contemptuous attitude toward developmental explanations, often surpassing that of most Modern Synthesis biologists, which is perhaps more aptly characterized as indifference; see Brown (2008) and Surbey (2008) for two good examples of the former.

[4] On the contrast between externalist and internalist approaches, as well as on the history of the debate, see Alberch (1991: 25–28).

1980). This fact was directly related to the increasing popularity of the conception of evolution fostered by the Modern Synthesis as a series of changes in the patterns of allelic distributions within the members of the same population, which justified the exclusion of the path leading from genes to adult features from evolutionary explanations—see Amundson (2005: Part I), Arthur (2004: Ch.3), and West-Eberhard (2003: Ch.1). The rationale underlying this position was that if evolution obtains by differential survival, favoring those organisms exhibiting traits that guarantee higher rates of reproductive success, then natural selection just needs to scrutinize the allelic variants that according to the Mendelian-Morganian idealization correlate with these traits within the gene pool of a population. Besides, as allelic variants are randomly and unrestrictedly introduced in particular genomic positions—the phenomenon known as "point mutation," it follows that natural selection is a self-sufficient creative force, conferring forms onto organisms in response to the environmental pressures operating on them. According to Amundson's (2005: 175) insightful analysis, this "black-boxing" of development was but a logical consequence of these ideas, held by the mainstream evolutionary thinking of the last century—see also Reid (2007: Ch.1), as well as Gould (1977) and Bonner (1982).

6.1.1 Evo Devo: Common background and alternative views

For several reasons, development cannot be dismissed from evolution as straightforwardly as the preceding argument seems to suggest. Curiously enough, some of these reasons are even congenial with other basic tenets of the Modern Synthesis. For example, it is clear that the simple fact of reaching the adult state is, for evolutionary concerns, as important as being an adult optimally fitted to overcoming environmental aggressions. In other words: It doesn't matter how hopefully fitted an adult you are if, to begin with, you are not a hopeful adult. This implies that alternative developmental pathways can also be deemed targets of natural selection—as Gilbert (2003b: 3) aptly puts it: "Every animal has to function as it builds itself." So, a course of development that makes the organism more robust from the start or, for example, allows an earlier emergence of certain key features, is probably to be selected instead of other alternative paths, with organisms following such a course flourishing and proliferating within the population. This is—we think—an idea that can be added "without tears" to the agenda of synthesis-oriented approaches,[5] in that it only requires conceptualizing a single developmental pathway as a specific phenotypic character, which correlates with certain genomic positions in the same simple and

---

[5] Waddington's (1957) concept of "canalization" can be seen as a pioneering formulation of this stance, in that it establishes that developmental paths become stabilized and strengthened by continuous exposure to the rigors of natural selection. Biases or constrictions on development can thus be explained as a direct consequence of standard Darwinian selection—for some comments on this matter, see Maynard Smith et al. (1985: 270).

linear fashion as other features of the phenotype as envisioned by the Modern Synthesis. Actually, this is the theoretical direction taken by a particular trend within Evo Devo, which relies on the assumption that evolution is mainly due to point mutations affecting genomic positions in charge of the regulation of genetic activity during development. This view—accessibly introduced in Carroll (2005)—can safely be judged a constructive enlargement of the scope of the strictly gene-centric view of the Modern Synthesis (Linde Medina 2010; Pigliucci and Müller 2010). An illustrative piece of this orientation is Gilbert et al. (1996), for they argue that the new synthesis of evolutionary and developmental biology has been provided by new findings from developmental genetics—namely, the developmental actions of certain (regulatory) genes and the homologies of genes and their domains of expression. Another—and particularly restrictive—version of this view is that defended in Carroll (2008), where the contention is made that evolutionary changes are largely due to mutations in the *cis*-regulatory sequences of developmental regulatory loci and their target genes. As development, according to this point of view, reduces to (genetically regulated) changes in gene expression,[6] a conceptualization of evolution as changes in gene frequencies continues to be valid within this general framework, albeit paying special attention now to certain developmentally critical genes—i.e., introducing a program of "population genetics of regulatory genes,"[7] in Gilbert et al.'s (1996: 368) illuminating expression.[8]

Such an extended-Modern Synthesis model is nevertheless explicitly rejected by a number of Evo Devo practitioners, who object to the simplistic view of development based on the linearity of gene–phenotype correlations and to the "monopoly" of the gene as a unit of inheritance—or "evolutionary currency."[9] In more positive terms, the common ground for biologists sharing this non-gene-centric orientation is (i) that factors other than genes—such as cellular products, mechanisms of cellular adhesion and communication, intermediate phenotypic states, environmental inputs, behavioral practices, and so on—have an important role in developmental processes; and (ii) that inasmuch as they exhibit individual variation, subject to one or another form of intergenerational transmission, they have an evolutionary potential to the

---

[6] In Carroll's (2008: 30) words: "Given that development is controlled by GRNs [gene regulatory networks], it follows that the evolution of development and form is due to changes within GRNs."
[7] Or, in Carroll's (2008: 34) own formulation: "Population genetics of CRE [*cis*-regulatory elements] variation and divergence."
[8] See also Gilbert (2003c: 350), where he presents Evo Devo as essentially a framework in which population genetics is "supplemented with or complemented by" developmental genetics.
[9] As aptly put by Rasskin-Gutman (2009: 75), within this reductionist molecular framework "it is evident that the black box between genes and phenotypes remains untouched." Similar critical stances can be found in Pigliucci (2007), Linde Medina (2010), and Benítez-Burraco and Longa (2010), among other places.

same extent as genes do.¹⁰ Hence, according to this general point of view, the Mendelian-Morganian idealization, justifying the evolutionary irrelevance of developmental phenomena, is clearly untenable. According to Müller's (2008: 17) summary:

> The paradigm of the synthesis was based on the correlation of phenotypic character variation with statistical changes of gene frequencies in populations. Adaptive change as population genetic event was the explanandum. The paradigm of evo-devo, by contrast, represents a causal-mechanistic approach towards the explanation of phenotypic change in evolution. Here the evolutionary alteration of developmental parameters (gene, cell and tissue properties, and their interactions) and their effects on phenotypic evolution are the explananda, whether adaptive or not.

This brief sketch of Evo Devo's response to the central tenets of the Modern Synthesis—clearly, a decreasingly influential position within contemporary biology—[11] is also probably helpful to convey the idea of what Evo Devo is and is not: Evo Devo is not a theory; it is at best a theoretical trend and above all an umbrella term under which different specific positions fall more or less comfortably—"No unified theory of Evo-Devo exists," in the words of Hall and Olson (2003b: xv); see also Benítez-Burraco and Longa (2010). Differences depend on how development is disentangled in terms of the number and nature of the factors involved and the complexity of the interactions they hold—see Robert et al. (2001) and, especially, Robert (2004). Thus Evo Devo projects range from those that preserve a deterministic idea of development as the unfolding of a "genetic program" almost intact (Carroll 2005, 2008) to those that postulate a "causal democracy" of sorts (Oyama 2000b), in which very disparate factors conspire to bring about reliable ontogenetic outcomes through intricate and contingent interactions (Griffiths and Gray 1994; Oyama 2000a; Oyama et al. 2001; Johnston and Edwards 2002). In between, some models extend factors beyond genes while maintaining the "program" metaphor (Keller 2000; Moore 2001), while other models depart from the "adultcentric" (Minelli 2003) view of the "program," but still maintaining certain limitations on the type and number of causal factors on development (Jablonka and Lamb 2005).¹² As a consequence, it seems difficult to understand Evo Devo—where updated versions of preformationist and epigeneticist views on development and of neo-Darwinian

---

[10] In Hall and Olson's (2003b: xiv) words: "Phenotypes and the processes that produce them are subject to selection; cells, embryos, and modifications of genetic and developmental processes are as much the raw material of evolution as are genes and mutations."

[11] But not in psychology, as attested by the growing body of literature in the field of evolutionary psychology—see above for some representative references, and Fodor and Piattelli-Palmarini (2010) for a recent discussion.

[12] See Gilbert (2003c), for a critical response to all these non gene-centered approaches.

and neo-Lamarckian views on inheritance coexist—as something more concrete than a generic vindication of the role of development in evolution.[13]

In any event, there is a group of topics that can be considered highly representative of the Evo Devo trend around which a certain degree of consensus exists irrespective of the specific orientation of particular workers. We think that the following four are worthy of special mention:

1. Some developmental pathways are extremely conservative, in the sense that they are manifested, with minor modifications, across very distant taxa. The recurrent use of these pathways within very different developmental contexts is a very significant feature of evolution;[14]
2. Modifications of developmental pathways, or what Arthur (2000, 2004, 2011) calls "developmental reprogramming" or "repatterning," obey a limited—and in itself constraining—inventory of descriptive categories: Displacements of onset and offset points, modifications of the rates of growth, alterations of the terminal state, changes in the plan of execution, and so on—see, among other sources, de Beer (1940), Gould (1977: Part II), Parker (2000), and Alba (2002) in addition to Arthur's references above;
3. Rigid developmental pathways are not expected outcomes of evolutionary change, as this would preclude innovation as an answer to strong environmental aggression or to the need of internal adjustment. So development itself likely evolves the capacity to react in the face of perturbations, a dimension of developmental systems customarily referred to as "evolvability"—see Hendrikse et al. (2007). This is not at odds with their "robustness," a complementary property referring to the capacity of reliably producing viable outcomes in spite of perturbations affecting internal or external factors on development—see Wagner (2005); and
4. The persistence, restrictiveness, and flexibility of developmental pathways and patterns of change are causally active factors that prevent us from taking the pressure of the environment as the only force governing evolution: (1)–(3) must thus be acknowledged with a creative character similar—if not superior—to that of natural selection.

---

[13] Or, in Müller's (2008) words, as a "discipline" harboring rather disparate theoretical positions. In point of fact, however, the Society of Integrative Comparative Biology (SICB) validated Evo Devo as an independent research area in 1999—see Goodman and Coughlin (2000) and Hall (2003) for some further comments.

[14] This is a very old idea, probably due to Aristotle. It is also one of the defining notions of nineteenth century (pre- and post-Darwinian) embryology, especially for the defenders of the idea that ontogeny recapitulates phylogeny—see Gould (1977: Part I) for a detailed historical account. It is fair to say that Evo Devo has rescued and updated this idea. Shubin (2008) contains an interesting and accessible revision of the similarities of the genetic and developmental background across the species.

These statements do not entail that natural selection is completely neglected by Evo-Devoists, who in general terms have just opened the issue concerning the extent to which the Darwinian mechanism can be deemed creative, now taking into consideration the constraints imposed by development. To be sure, Evo Devo is also particularly open with respect to this matter with positions ranging from classical "genic selectionism" (Carroll 2005) to a radical redefinition of selection as a "filter," rather than a creative mechanism (Goodwin 1994; Kauffman 1995).

6.1.2 "Morphological evolution" and related concepts

Alberch—in particular (1980), (1982), (1989), and (1991)—originated the concept of "morphological evolution" aimed at explaining the phylogeny of organic designs on the very same basis as modern Evo Devo thinking. Changes in the parameters underlying the development of organisms are the main source of the evolutionary processes capable of introducing novelties in nature.[15] He also contended that developmental systems possess the properties of complex dynamic systems (Thelen and Smith 1994; Kelso 1995; and Webster and Goodwin 1996, for an earlier application of these ideas to biology), in which intricate interactions between genetic and non-genetic factors relate non-linearly with morphological outputs. Such non-linearity means that once certain critical thresholds are reached, small perturbations of any of the morphogenetic parameters of the system—for example, the kinetic activity of cellular diffusion, the viscoelastic properties of cellular matrices, the mitotic rates, and so on—are capable of bringing about wide-ranging consequences in development and, ultimately, in the evolution of entire lineages of organisms. In the meantime, changes may continuously be going through without major—if at all—somatic manifestations. In this connection, De Renzi et al. (1999: 625–626) have pointed out that Alberch anticipated the application of contemporary complex systems theory to organic development through his conception of developmental systems as dynamic non-linear systems—see in particular Oster and Alberch (1982), and Alberch (1982, 1991); see also Kauffman (1993), which includes a note of personal acknowledgment and several references to Alberch's work.[16]

Alberch also shared with most Evo Devo theoreticians the conviction that the most common effects of those perturbations have to do with the timing and/or rates of growth of the developing structures, which as a result can diverge radically from closely related ones (Gould 1977). "Heterochrony" was thus the main pattern of organism-level change referred to by Alberch to explain the origins of most

---

[15] On the life and works of Pere Alberch, see Rasskin-Gutman and De Renzi (2009) and Reiss et al. (2009).

[16] As a matter of fact, at the time of his death Alberch was working on the project of formalizing his ideas in terms of chaos and complexity theory in a book entitled *An Introduction to Chaos Theory and Complexity With Special Emphasis on Biological Sciences*—L. Nuño de la Rosa (pers. comm., 11 March 2008).

evolutionary novelties (Alberch et al. 1979; Alberch and Alberch 1981). Regarding this matter, Alberch adopted a somewhat critical stance and rejected the prevailing view on heterochrony, based on the comparison between phenotypic states of organisms at certain stages of their developmental sequences. He contended that within this view any kind of mutational phenomenon could bring about a heterochronic outcome inasmuch as its phenotypic manifestation could be deemed similar to a different developmental stage of one or another closely related organism. In Alberch's opinion however, the concept of heterochrony should be applied not to developmental states but to rules and interactions acting on developmental processes and it should specifically refer to changes in their timing and rates, irrespective of how they translate into the phenotype's makeup (Alberch 1985 and Alberch and Blanco 1996; see De Renzi 2009, Etxeberria and Nuño de la Rosa 2009, and Reiss et al. 2009, for some comments).[17] He thus settled the definition of the concept as it is currently used in Evo Devo (Smith 2001, 2002).

A more distinctive aspect of Alberch's model is the idea that systems of interactions underlying developmental sequences are rather stable and that changes in these sequences are mostly due to modifications in the values of one or another of the morphogenetic parameters composing the system (Alberch 1989, 1991). In the study of complex dynamic systems, the concept of "control parameter" refers to the systemic component whose perturbations correlate with the emergence of new morphologies—a new pattern in the surface of a chemical solution, a new embryological state, a new form of behavior, and so on (Thelen and Smith 1994: 63-64). A control parameter is not a central agent in the causation of phenotypic variation, in that the effects of its perturbations are not immediately reflected on the morphological outputs but on the other morphological parameters (Thelen and Smith 1994: 112; Kelso 1995: 7). So, within this model the emergence of new variants is always a joint function of the system as a whole. The idea of control parameter basically introduces the possibility of pinpointing a single parameter of the system as the starting point of the chain of reactions leading to new morphologies.

For our own purposes, however, the most relevant aspect of Alberch's proposals is the contention that developmental systems foreshadow the scope of the attainable phenotypes, as well as the trajectories leading from one phenotypic state to another. It is the concept of "parametric space" that in Alberch's framework is in charge of theoretically representing the finite and discrete set of the possible outcomes of any developmental system (Alberch 1989, 1991). The main properties of parametric spaces are summed up in the following paragraphs—see figure 6.1 as a point of reference.

A parametric space is a finite set of discrete or discontinuous phenotypes sharing a common underlying system of morphogenetic parameters (x and y in figure 6.1—

---

[17] Müller (2008: 9) notes, however, that distinguishing between instances of one or the other kind of phenomenon can be difficult.

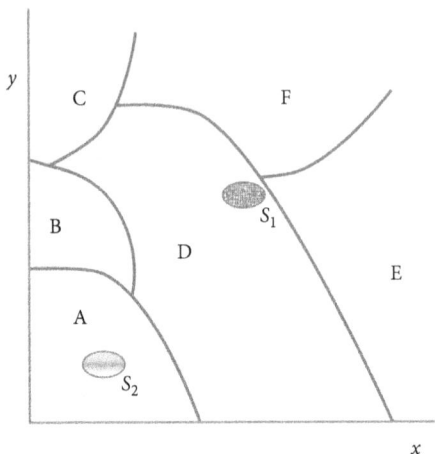

FIGURE 6.1 **Guess where your phenotype is**
Developmental systems describe discontinuous spaces of species-typical phenotypes. Phenotypes (A–F) and species (s_1, s_2) differ in terms of relative stability within their particular morphospace and phenotype respectively.

obviously a very rough idealization). The discontinuous character of phenotypic variation is captured in the figure by the regions labeled with capital letters. Each phenotype has a particular probability of coming into being, represented by the extension that it occupies—D is thus the most probable phenotype, while B the most improbable one. Moreover, each phenotype is also characterized by the relative probability of transforming itself into one or another of the neighboring phenotypes. In figure 6.1 this aspect is represented by the length of the line demarcating different phenotypes—A has a strong probability of turning into D, a low probability of turning into B, and no probability at all of turning into any of the remaining phenotypes.

Species—and also species' traits—are represented in figure 6.1 by means of the oval shapes s_1 and s_2. From a populational point of view, the model incorporates the following contentions: First, every species falls into one or another phenotype—s_1 belongs to phenotype D, whereas s_2 fits into phenotype A; second, the morphological stability of a species is a function of both (i) the probability of its phenotype—s_1 is, in principle, a more stable population than s_2— and (ii) its proximity to a point of bifurcation to other phenotypes—s_1 is thus a rather unstable population within its phenotype, given its vicinity to the bifurcation leading to E and F; finally, both the proximity to a point of bifurcation and the relative propensity of its own phenotype to transform into one or another phenotype put a certain population on the edge of undergoing a radical morphological reorganization—s_1, for instance, has a high propensity of acquiring the properties of phenotype E. "Bifurcation" is thus another key concept of Alberch's framework (Alberch 1982; Oster and Alberch 1982). It identifies

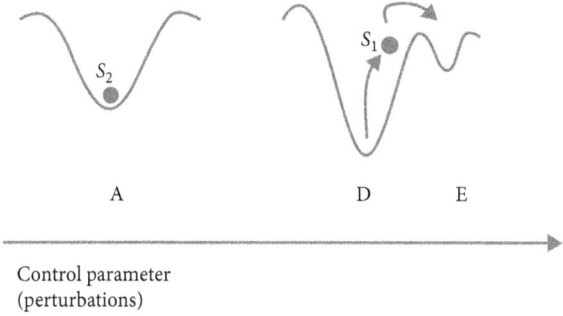

FIGURE 6.2 **Guess where your phenotype is (from another view)**
Species-typical morphologies (like that of s_2) are said to be deeply embedded within a well of attraction when they are highly stable within its phenotype (A). However, a history of continuous perturbations may put a species (s_1) at the edge of a well (D), until they abruptly make the species shift into another well (E). From that moment on the species will exhibit a distinctly new morphology.

points within a parametric space in which a minimal perturbation is capable of bringing about qualitatively new morphologies. Using the jargon of dynamic systems theory—whose familiarity with Alberch's model has been underlined before, certain populations (e.g., s_2) are deeply and stably accommodated in a "well of attraction" as depicted in figure 6.2, while others (e.g., s_1) are sitting at the edge of a well—or close to "phase shift"—due to their continuing exposure to a certain kind of perturbation, a situation that makes them highly susceptible to entering into a different well of attraction.

In any event, Alberch's own favorite model for representing the idea of bifurcation more vividly within a parametric space was Waddington's epigenetic landscapes (Alberch 1982: 324). As a matter of fact, epigenetic landscapes—figure 6.3—are just wells of attraction, if given a third dimension and thus transformed into grooves: Each groove describes a developmental trajectory, which at certain points connects to alternative, less probable trajectories, into which development can nevertheless enter if somehow perturbed.

Alberch's spaces are also congenial with Waddington's landscapes in that complex systems of interactions—or "chemical tendencies"—underlie the latter, ultimately reducing to the genes on which those tendencies are grounded (figure 6.4) in a way similar to that in which complex intra- and supra-cellular parameters constitute the axes defining the former. In Alberch's view these parameters are hierarchically organized with gene activity also making up the ground level although with an equally relevant role from a causal point of view.

Take now, as an illustration in point, the morphogenesis of the dermal organs—see Odell et al. (1981) and Oster and Alberch (1982). Skin tissue may form the basis of very disparate structures—hair, salivary glands, teeth, feathers, scales, limbs, and carapaces, among others—that are discontinuous from species to species, in the sense

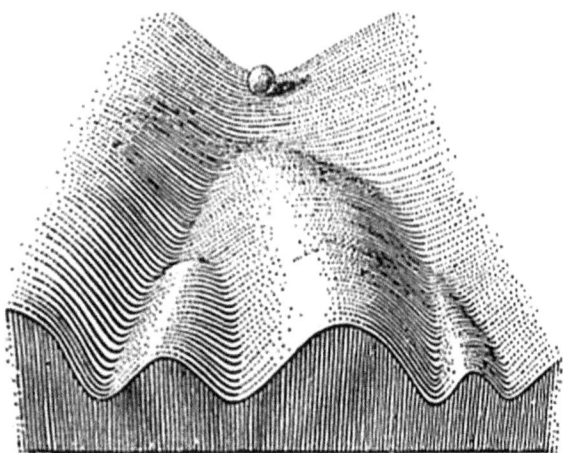

FIGURE 6.3 **Guess how your phenotype will slide**
Species tend to reliably follow certain developmental paths instead of alternative accessible ones, just as balls tend to follow deeper grooves instead of shallower yet connected ones. Perturbations may nonetheless make them enter into different paths. Small perturbations can be enough at certain critical points to change the developmental fate of a species.

Original image from Waddington (1957: 29).

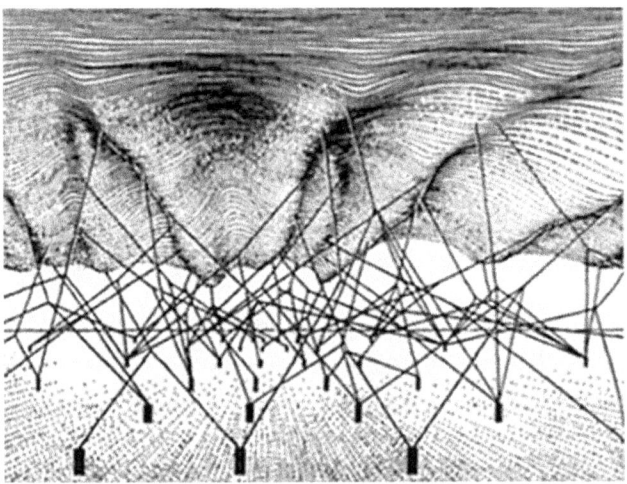

FIGURE 6.4 **Guess how your phenotype will slide (a view from below)**
Chemical tendencies and ultimately genes are the determinants of developmental paths—just as strings and pegs may underlie the grooves of an artificial slope. Gene activity and its chemical consequences can be perturbed as to change the developmental fate of an organism—just as the shape of a slope changes by modifying the underlying system of strings and pegs.

Original image from Waddington (1957: 36).

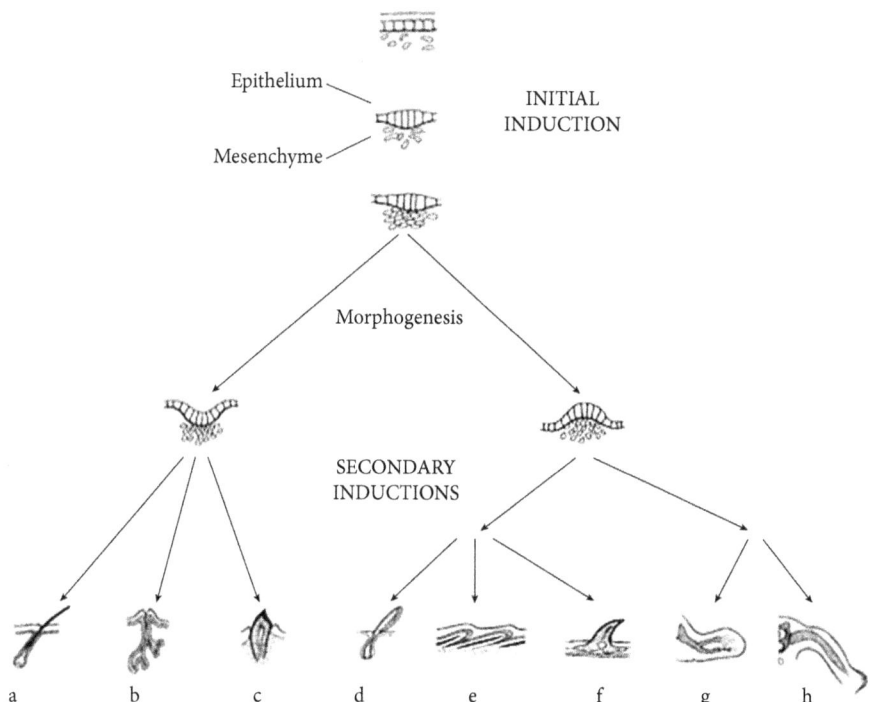

FIGURE 6.5 "I've Got You under My Skin"
Slight divergences in the epithelium's thickness and the mesenchymal tissue's concentration at different developmental stages bring about qualitatively different dermal organs (hair, glands, teeth, feathers, scales, carapaces, etc.) from the same developmental system. The figure, adapted from Alberch (1989: 47), illustrates the morphospace concept at the organ level.
Caption: song by Cole Porter (1936).

that no structure can be said to be a transition form between any other two structures. They all originate however in equivalent inductive processes having to do with the thickness of the epithelium and the concentration of the mesenchymal tissue—figure 6.5. Depending on the elasticity and the strength obtained in each case, the epithelium can (i) invaginate, giving place to dermal phenotypes with hair, glands, or teeth, or (ii) evaginate, giving place to dermal phenotypes with feathers, scales, or carapaces. Thus, each structure is only accessible through perturbations acting upon a particular phenotypic stage in the course of development, according to an evolved species-typical sequence.

From an evolutionary perspective, an important corollary of this model is that the geometry of parametric spaces works as a very strong constraining force, capable of counteracting that of natural selection. As is also the case with other Evo Devo-oriented approaches (see section 6.1.1), this does not discredit the role of selection in the evolution of organic designs, but—as explicitly stated by Alberch (Alberch 1980: 664, 1982: 318–319, 1989: 46–48, 1991: 16; Oster and Alberch 1982: 45)—it redefines selection as a filtering mechanism acting on designs created independently and whose variation is also independently biased.

The next section is devoted to exploring how this general framework can be extended to the realm of cognition.

6.2 Morphological Evolution and Computational Phenotypes

> Where is my mind?
>
> Pixies, 1988

In the previous chapter we introduced the idea that the Chomskyan family of formal languages—traditionally ordered in a hierarchical series of increasing complexity, as in figure 5.2, can be interpreted as a family of Owenian archetypes, corresponding to qualitatively different regimes of computational activity—or "computypes." We are now in the position to present the complementary idea that the hierarchy as a whole can be thought of as an Alberchian parametric space—or "developmental morphospace." Remember that we have previously and independently justified the following claims: (1) The "activity" level is a legitimate one for applying the comparative method in biology; (2) "computation" refers to a distinct type of organic activity; and (3) diversity in this particular domain shows the hallmark of discontinuous variation. So we believe that our proposal in this section relies on solid conceptual grounds. For us, this is an important section because it traces a theoretical framework in which the Chomskyan types—now archetypes—play a crucial role in an explanatory scheme concerning the evolution of CSs, a family of cognitive systems subserving skills as disparate as birdsong or language—plus others to be presented in the next chapter, so they get full legitimacy as natural kinds. The following subsection is devoted to this particular topic; next, in a separate subsection, we advance some proposals aimed at explaining the evolutionary transition leading to CS_{HUMAN} as the outcome of a coincident collection of heterochronic phenomena in the evolution of the human brain and, particularly, in the memory/cortical component of the organ of computation. Partial as these proposals will be, we believe that they are important in that they demonstrate that giving names to some of the developmental parameters describing the morphospace of possible computational phenotypes is not beyond the reach of contemporary research. They are also crucial, or so we believe, in showing that the idea that the same organ of computation exists in different species—differences in anatomy or activity notwithstanding —is empirically testable. As knowledge concerning the developmental underpinnings of this putative organ accumulates in the near future, the claim that it comprises a collection of homologues linked by a pattern of heterochronic diversification will be clearly exposed to the test of fact.

6.2.1 The computational morphospace

The Chomsky Hierarchy comprises a set of discrete and easily identifiable models of computation, two properties also exhibited by phenotypes within Alberch's morphospaces. This is the main inspiration underlying our idea of translating the former into a space of computational phenotypes—or "computypes," possessing the same pattern of discontinuous distribution of phenotypic variation of the latter. Computypes are not simply quantitative variants of each other; instead, each particular computype implies the introduction of a genuine novelty, in a qualitative sense: $CompT_2$ introduces, relatively to $CompT_3$, the equivalent of a push-down stack; and $CompT_1$ introduces, relatively to $CompT_2$, not just the equivalent of an enhanced push-down stack, but of a system of embedded push-down stacks. Obviously, the unlimited space and time resources of $CompT_0$ also represent a qualitative improvement relatively to any other system of computation, but in this case it does not correspond to a naturally realizable system. Under this interpretation, there is quantitative or continuous variation within each particular computype—corresponding perhaps to the levels of embedding of the memory stacks in the case of $CompT_1$, or to the degree of memory resolution in the case of $CompT_2$, but computypes are qualitatively or discontinuously different from each other. Thus, an organism showing a particular computype—say, $CompT_2$—is comparable to a phenotype showing a particular dermal organ—say, teeth (figure 6.5), while different organisms sharing $CompT_2$ are comparable to phenotypes showing different types of dentition.

A developmental morphospace captures a certain pattern of phenotypic variation. Besides, it also captures the idea that variants located in one or another position within this space, whether belonging to the same phenotype or not, are deeply homologous, in that they are all outcomes of a unique system of factors and constrictions on development. In the next section, we offer some—assumedly sketchy—ideas regarding such a system in the particular case of the organ of computation. For the time being, we can recur to Alberch's idealization, introduced in the previous section, and abstractly represent the morphogenetic parameters of the system by means of two axes describing a flat, two-dimensional space, as in figure 6.1 above. However, this figure needs some minor adjustments, as we have no information at all concerning each computype's probability of transforming itself into one or another computype. As an alternative, we can represent computypes as occupying unconnected patches within the morphospace—each patch dependent on the specific values of the morphogenetic parameters, with arrows standing for the developmental pathways leading to an alternative computype (figure 6.6).[18] A shortcoming of this

---

[18] Alberch himself used this convention in some of his works (1980: 655, 1982: 316, and 1989: 25), as an alternative to that of figure 6.1 above.

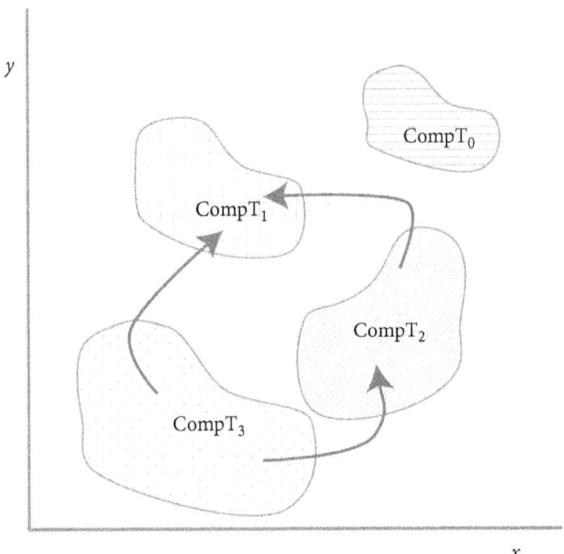

FIGURE 6.6 **Guess where your computype is**
The computational morphospace depicts a discontinuous space of species-typical computypes. Species may evolutionarily attain a new, eventually more complex CS due to perturbations affecting the parameters underlying their development.

convention is that the size of a patch represents the probability of the corresponding computype relatively to other computypes, of which no reliable information exists. However, we can ignore the exact size attributed to each computype in figure 6.6, and just assign each one a relative probability in inverse proportion to its complexity. This is an intuitive contention for which we have no empirical support. In any event, the point of figure 6.6 would remain untouched, were this particular intuition disconfirmed. An advantage of this figure is, on the other hand, that it emphasizes the discrete character of phenotypes and the discontinuity of variation—consequently, transitions from one to another phenotype are "jumps."

Note that this representation foreshadows a pathway leading from each computype to all hierarchically higher computypes, but not the other way around. This captures the idea that, depending on the nature of perturbations and the particular factors affected by them, outcomes can profoundly diverge from the common phenotype of a population; it also captures a departure from a strictly Darwinian model of evolution, where the raw material on which natural selection acts are always minute deviations from previously existing phenotypes. Thus, interpreting the Chomsky Hierarchy as a computational morphospace along the lines depicted in this section does not force us to envision the evolution of organs of computation as an orderly process following the different levels of computational complexity, as in a *scala naturae* of sorts—again, our keyword is "jump."[19] This is, by the way, an aspect of our model that is in complete agreement with Alberch's dynamic proposal concerning heterochronic phenomena—outlined in the previous section, according to which recapitulatory sequences are not to be expected in most developmental processes (Alberch 1985: 51). In the particular case of CS_{HUMAN}—located within the CompT1 area, no strong piece of evidence exists in support of the idea that it directly evolved from a CompT2 or rather from a CompT3 phenotype. However, if we follow Hauser et al.'s (2002) claim concerning the scarcity of CompT2s in nature, and we further base on this consideration that CompT3-to-CompT1 is more probable a transition than CompT2-to-CompT1 in order to attain CS_{HUMAN}, the good news is that our model is perfectly fit to deal with this evolutionary scenario.

Finally, another important detail of figure 6.6 is that there is no arrow pointing toward CompT0, which is our way of capturing that there may not exist any possible developmental path leading to it. Contrary to what it may seem at first glance, this is not problematic. As pointed out by Rasskin-Gutman (2005: 214–215), we must distinguish between "theoretical morphospaces," including possible (both actual and potential) and impossible phenotypes like the one in figure 6.6, and "empirical morphospaces," excluding the latter—i.e., figure 6.6 minus CompT0. Their exclusion is justified by natural limitations in the parametric factors involved: In our case, no factor can be perturbed as to produce a device with an unlimited potential.

Our naturalized version of the Chomsky Hierarchy, transformed into a developmental morphospace, is now complete. With this picture in mind, we can refer to the origins of CS_{HUMAN} as the evolutionary outcome attained at a critical point of a process of continuous perturbations affecting certain morphogenetic parameters of the system underlying the development of the organ of computation across species. In this connection, it is important to take into account that—as already noted by Bateson (1894: 13–17)—the fact that there exists discontinuity in variation does not contradict the idea of gradual change at the level of processes—see figure 6.2. This observation implies that, at a computational level of analysis, language can be deemed "discontinuous" relatively to any other CS-subserved skill (but see next chapter for some important qualifications of this tenet), while still be "continuous" at the level of the developmental processes leading to the corresponding computational phenotypes—which in turn implies that the "Continuity Paradox," as traditionally presented (Chomsky 1968; Bickerton 1990), is probably trivial. Thus, it seems to us that a more suitable terminology in order to characterize the system of computation subserving CCC_{HUMAN}—the system upon which language, among

---

[19] In addition, even though this is not explicitly represented in the figure, nothing prevents a jump "backwards," from a more complex to a simpler computype, and thus capturing the possibility of evolutionary events of reduction in computational power perhaps parallel to the putative episodes in brain reduction proposed for some metazoan lineages (Striedter 2005; Schmidt-Rhaesa 2007).

other skills, is parasitic—is the concept of "critical-point emergence," as defined by Reid (2007: 398): A new or enhanced function that appears in the manner of a saltation in a continuous series of changes—e.g., wing enlarges to a point where lift exceeds drag and gravity, gliding to true flight (Reid's example). As a matter of fact, Reid (2007: 131) himself, in passing, mentions language as a likely example of the idea:

> My own opinion is that human linguistic ability is a qualitatively critical-point emergence correlated with cerebral expansion and reorganization, one of the major anatomical emergences of fetalization. It is not a product of quantitative accumulation of little bits and pieces that have conveyed sufficient advantage to have gradually produced human speech.

In the next section we advance some specific proposals about the developmental processes paving the way for the emergence of CS_{HUMAN}.

6.2.2 Heterochrony and constraints in brain evolution

It is time to turn our attention back to the anatomical description of CS_{HUMAN} that we suggested in Chapter 5. We want to add now to this suggestion some specific ideas aimed at explaining how such an anatomical structure has achieved its computational power in the course of human evolution. Two closely intertwined proposals articulate our main thesis on this matter, both directly inspired by the model of morphological evolution introduced in the previous section:

1. CS_{HUMAN} has evolutionarily attained its characteristic computational power—i.e., its computype—as a consequence of a set of heterochronic phenomena in the evolution of the human brain; and
2. The outcome of this process has been very rigidly constrained by properties of the developmental system on which the heterochronic perturbations at issue have acted.

As for thesis (1), heterochrony—according to Gould's classic formulation (1977: 4)—occurs when characters already present in a particular organic design undergo changes in developmental timing[20] or intensity—i.e., dosage of cellular products. As we contend that CS is an extremely common organic system—extending, at a minimum, to the whole vertebrates phylum (but see Chapter 8 for additional considerations on this particular issue), a heterochronic model of evolution seems specially fitted to the case.

---

[20] Changes in the location of structures—a phenomenon specifically known as "heterotopy," also belong to the family of repatterning phenomena (Zelditch and Fink 1996). However, for the sake of simplicity—but also for the "primacy of time" (Minelli 2003) in development, "time" can safely be used as a covering term in the description of changes of developmental patterns, but see Arthur (2011) for a detailed overview of the different types of cases of developmental repatterning.

TABLE 6.1. Pick your favorite heterochrony

1. PAEDOMORPHOSIS. Underdevelopment or terminal truncation	2. PERAMORPHOSIS. Overdevelopment or terminal extension
1(a) NEOTENY. Reduced rate of development	*2(a)* ACCELERATION. Increased rate of development
1(b) PROGENESIS. Earlier offset	*2(b)* HYPERMORPHOSIS. Delayed offset
1(c) POSTDISPLACEMENT. Delayed onset	*2(c)* PREDISPLACEMENT. Earlier onset

Homologous organic designs are linked by heterochronic patterns of diversification when they depart from each other for reasons having to do with developmental timing or intensity. Based on McKinney (2000) and Parker and McKinney (1999)—see also Gould (1977); McKinney and McNamara (1991); and Alba (2002).

Table 6.1 offers a typology of heterochronic processes—somewhat simplified, but sufficient for our purposes.

The field of developmental neurobiology offers a solid body of literature pointing to the conclusion that the nervous systems of closely related species differ—sometimes strongly—according to patterns of variation easily describable using heterochronic categories like those defined in table 6.1. As some of these observations relate to the nervous subsystem that we identify as the physical realization of CS and, besides, they refer to the development of mammalian and, more specifically primate brain structures, we offer in the following paragraphs a brief review of this comparative data.

That both larger brain size and the connectivity rate among its components—especially but not exclusively at the cellular level—may obtain from alterations at very early stages of embryonic development, is by now a well-established fact. In this connection, Kaskan and Finlay (2001) point out that an increased rate of production of precursor cells or an extension of cytogenesis during embryonic development may result in significant differences in brain size. These authors define "cytogenesis" as the period spanning from the point where production of precursor cells of some structure starts to the point where a maximum of cell division obtains and the resulting cells do not divide but "migrate" to the forming structure (Kaskan and Finlay 2001: 17). Moreover, they note that in the development of different brain structures from the neural tube of mammals one observes clear displacements affecting the highest point at which such asymmetric terminal division is reached. Crucially, they also claim (i) that the resulting neural population grows exponentially with respect to the displacement of this peak; and (ii) that a greater displacement is systematically observed in the formation of the cortex—C in figure 6.7—compared to other structures such as the spinal cord or the basal ganglia—respectively, A and B in figure 6.7; from Kaskan and Finlay (2001: 20). This is so because, according to Finlay

128 Sergio Balari and Guillermo Lorenzo

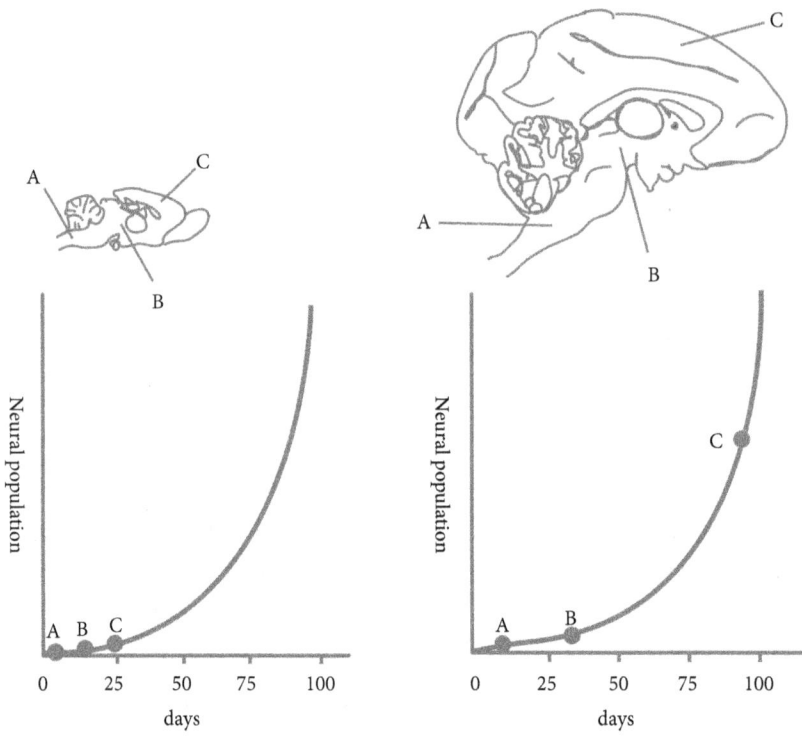

FIGURE 6.7 Climbing Mount Cytogenesis: Guess where you are
Homologous brain structures of rodents (left) and primates (right)—namely, the spinal cord (A), the basal ganglia (B), and the cortex (C)—differ according to a combined pattern of peramorphosis—increased rate of production of precursor cells (acceleration) and delayed offset of cytogenesis and asymmetric cell division (hypermorphosis), notably intense in the case of the last-named structure. Given the exponential character of the growth of neural populations, the primate cortex gets disproportionate dimensions relatively to the other non-cortical structures.
Based on Kaskan and Finlay (2001).

and Darlington (1995), these nervous structures obey a pattern of concerted development, the schedule of which is furthermore subject to a general principle of evolutionary conservativeness. So, evolutionarily speaking, the schedule is expected to stretch or to squeeze, but not to undergo rearrangements; and, if it stretches, displacements are expected to occur concertedly and in a proportional fashion, giving place—as commented—to an exponential growth of relevant structures. Even if the mechanistic basis of this phenomenon is unknown, it fits well with existing comparative data (Striedter 2005).

From a comparative point of view, two extremely important consequences of Finlay and collaborators' findings are (i) that a few days of displacement can bring about a very strong contrast in the cortical mass attained by different species—as in the case of rodents (figure 6.7, left) vs primates (figure 6.7, right); and (ii) that a meaningful contrast follows across mammalian species between their rather uniform subcortical structures and their much more diversified cortical structures. We connect in a while this latter observation with our thesis (2) above.

Another important comparative quantification concerning the development of the mammalian cortical structure is offered in Rakic and Kornack (2001). They point out that the phase of asymmetric cell division yielding to neural cells—that is, the phase to which peak Kaskan and Finlay (2001) refer—starts in monkeys some four weeks later than in mice, which entails an extension of the anteceding period of symmetric cell division, where the majority of neural precursors are produced. According to the data presented by these authors, the population of neurons in monkeys doubles that of mice, as a result of this displacement at the onset of a characteristic phase of embryonic development. Moreover, as Rakic and Kornack (2001: 45–46) also point out, in the case of humans the onset of asymmetric cell division is displaced only a few days later than that of monkeys, but, given the exponential effect of such displacements on the production of neural precursors, the population of neurons in humans is estimated to be between some eight or sixteen times larger than that of monkeys.[21] In the light of this data, the authors conclude that the mutation of some regulatory gene—or a collection thereof—may have been responsible for the reorganization of the neo-cortex and of the cognitive and, ultimately, behavioral novelties associated with it (Rakic and Kornack 2001: 46). Table 6.2 shows an assortment of candidate genes related to brain size that, according to investigations conducted in the last few years, have undergone positive selection in the recent evolutionary history of humans. It is worth remembering at this point, however, that we are far from contending that any of these genes is directly responsible for the evolutionary emergence of language—actually, there does not seem to exist such a thing as "the gene (or genes) for language" (Benítez-Burraco 2009). According to our evolutionary model, they are more likely control parameters, whose perturbations may have triggered a chain of reactions extending to the whole system of development.

Other differences of the developmental sequence of the human brain, as compared to that of monkeys, are also known, and they coincidently point to congenial patterns of delay. For example, Parker and McKinney (1999) and McKinney (2000) contend (i) that fetal growth of the human brain is 25 days longer than that of monkeys, (ii) that myelinization of the neo-cortex—and, in particular, that of the frontal areas—is

---

[21] Both Kaskan and Finlay's (2001) and Rakic and Kornack's (2001) observations can jointly be comprised under Finlay and Darlington's (1995) "late equals large" rule. According to this rule, brain regions that mature relatively late become disproportionately large. See Striedter (2005) for some comments.

TABLE 6.2. Pick your favorite mutant

Gene	Function	Selection effects	References
MCPH1	Effects on brain size through an increase in the proliferation of cell death during neurogenesis	Strong positive selection (SPS) within the primate lineage that gave rise to Old World monkeys and higher apes	Gao et al. (1998); Hakem et al. (1996); Evans et al. (2004a); Wang and Su (2004); Evans et al. (2006)
ASPM	Maintenance of an asymmetrical division pattern of neuronal precursors	SPS after the human lineage branched from the evolutionary line leading to higher primates	Bond et al. (2002); Zhang (2003); Evans et al. (2004b); Woods (2004); Mekel-Brobrov et al. (2005); Yu et al. (2007)
CDK5RAP2	Unknown	SPS in the human lineage	Ching et al. (2000); Dorus et al. (2004)
CENPJ	Microtubule nucleation	Ditto	Hung et al. (2000); Hung et al. (2004); Dorus et al. (2004)
SHH	Regulation of the morphogenesis of the ventral region of the neural tube. Regulation of axonal growth and synaptogenic processes in some neuronal circuits	SPS within the primate group	Echelard et al. (1993); Roelink et al. (1994); Marigo et al. (1995); Roessler et al. (1996); Dorus et al. (2004); Bourikas et al. (2005); Salie et al. (2005)
CASP3	Activation of apoptosis during the multiplication of neuronal precursors	Unknown	Kuida et al. (1996)
ADCYAP1	Modulation of neuronal precursors' transition to differentiated neurons	SPS after the branching of the lines leading to chimpanzees and humans	Dicicco-Bloom et al. (1998); Mercer et al. (2004); Wang et al. (2005)

Many genes have recently been identified correlating with brain size in humans, so according to our anatomical and developmental claims they are reasonable candidates for being ingredients of the recipe for the computype evolutionarily attained by humans. See Balari et al. (in press) for further details and comments.

extended in humans until the age of 12, whereas in rhesus monkeys it lasts only 3.5 years, and (iii) that dendritic growth is extended in humans until the age of 20, well beyond that of any other of its kindred species. As all these phenomena have to do with factors affecting neural interconnectivity rates, their combined effect must no doubt have some far-reaching consequences in human cognitive development, as explicitly argued in Parker and McKinney's (1999) comprehensive study on primate comparative cognition. As a matter of fact, Langer (2000) observes that the comparison of cognitive development in monkeys, chimpanzees, and humans in areas where it is possible to directly apply a comparative method—as in physical and logico-mathematical intelligence—also reveals a generalized pattern of "overdevelopment" or "terminal extension" that matches the degree of evolutionary proximity relatively to the human species. As Langer himself emphasizes, this is an extremely important conclusion, because it is consistent with the existence of processes of overdevelopment at the level of brain structure affecting different aspects of brain anatomy—glial cell growth, axon myelinization, synaptogenesis, or dentritic growth (Langer 2000: 229).[22]

The above comments are also illustrative of our thesis (2), because they clearly speak of a highly constrained system underlying the development of different nervous structures in a concerted fashion (Finlay and Darlington 1995), instead of in a more mosaic-like manner (Barton and Harvey 2000). As pointed out by Kaskan and Finlay (2001: 27), the brain seems to be one of the few organs whose development appears to follow a set of rules affecting the organ as a whole. This means that the fate of a particular brain structure seems inescapably linked to that of the rest of the structures with which it establishes functional connections and, consequently, that the amount and type of variation upon which selection can act in brain evolution is drastically biased from the start.

Furthermore, the evolution of the nervous system seems to be constrained in other subtle ways,[23] as argued for example in Hofman (2001), where the claim is made that certain aspects of the *Bauplan* (Hofman's term) of the primate brain act as very rigid limiting constraints on its evolutionary potential. Hofman specifically points out that the exponential character of increments in the cortical mass as brains get bigger contrasts very strongly with that of other brain structures with a relevant role in regulating cortical activity—like the cerebellum or the basal ganglia, where growth follows a more linear pattern. As a consequence, the more the brain grows, the more disproportionate the relation is between its cortical and subcortical structures, with the latter eventually getting functionally inadequate for the kind of complex

---

[22] For the sake of accuracy, what Langer exactly observes is a generalized pattern of displacement in the offset of both aspects of intelligence as we get closer to the human species, as well as a pattern of displacement in the onset of logico-mathematical intelligence, which starts earlier in humans than in chimpanzees, and in chimpanzees than in monkeys—see Langer (2006) for a short but complete overview of his observations. But remember that according to Alberch's model, Langer's statements are not per se heterochronic observations—or they are not until causally connected to heterochronic phenomena at the level of the underlying developmental mechanisms. This means that both findings—i.e., the displacement of the onset and the displacement of the offset in the corresponding cognitive areas—could perfectly be due to a single heterochronic phenomenon—be it predisplacement or, more likely, hypermorphosis.

[23] Metabolic and energetic considerations aside, which nevertheless are important factors in limiting brain evolvability.

information processes the former would otherwise be able to undertake. An important evolutionary consequence follows from this observation, namely that even if the cortex is a structure which undergoes transformations much more promptly than other (subcortical) brain structures—Striedter (2005) for an overview, its evolvability is nonetheless strongly constrained by the archetypal organization of the organic system that it belongs to, given the conservativeness of subcortical brain components.[24] Again, all this purports to saying that the raw material for brains to evolve seems to be strongly biased before selection can be said to act upon it.

The fact that "most brain evolution is of the concerted kind"[25] (Striedter 2005: 169) thus points to the conclusion that most aspects of brain morphology and, consequently, of brain activity have not been naturally selected by standard Darwinian means. Not just because morphological variants are not randomly distributed among populations—but dramatically biased by the underlying developmental system; also because within a framework of concerted evolution of a rather complex structure, associated to many different cognitive specializations, no single function can be pinpointed as the advantageous one historically responsible for the evolution of a certain morphology. This is a well-known problem that, according to Fodor and Piattelli-Palmarini (2010), defeats the logic of adaptationism: As functions hitch-hike, no one can tell a single function apart as ultimately responsible for the evolutionary fate of a structure in the absence of counterfactuals showing which one would remain in the absence of all the others. This automatically implies that evolution, taking place in a world of contingency in which no counterfactuals de facto exist, cannot be scrutinized by the eyes of natural selection if structures obey patterns of concerted development involving an assortment of functions.[26] Thus, brain evolution, a structure in which functions massively hitch-hike—and necessarily so for developmental reasons—presents itself as one of the most challenging counterexamples to the logic of standard selectionism—a hypothesis we had no need of in this chapter, as Laplace would have put it.

---

[24] The basal ganglia have been so described in a number of comparative approaches; see Reiner et al. (1984) and Reiner (2010).

[25] This does not mean that brains are completely free from mosaic-like evolutionary processes. In any event, as pointed out by Striedter (2005: 157), such processes will more likely have to do with late episodes of brain development, in which developmental constraints somehow relax, without altering the essential aspects of brain architecture, and they most probably serve the fine-tuning of already existing crucial structures or connections. See next chapter for some ideas along these lines concerning the case of CCC_{HUMAN}.

[26] See also Gould and Lewontin (1979), as well as Huxley (1932) for some seminal ideas in the same direction. For some criticisms on this and other aspects of Fodor and Piattelli-Palmarini's controversial book, see Block and Kitcher (2010), Pigliucci (2010a), and Barrett (2011).

6.3 Conclusions

The fit of (i) our computational take on animal cognition, (ii) the physical interpretation of the organ of computation based on Lieberman's (2006) Basal Ganglia Grammar model, and (iii) the computational morphospace describing the developmental and evolutionary trajectories of this organ, should by now be clear. In the case of CS_{HUMAN}, continuous perturbations affecting the morphogenetic parameters that define the developmental system of CS—probably, but not necessarily, starting with point mutations in some key genes—drove the system to a "bifurcation point," from which some further—even if minor—perturbations were enough to locate the system into a "phase shift." As a consequence, the attainment of a qualitatively new regime of computational activity—CompT1—was reached—without necessarily going through a CompT2 stage, a detail that requires independent empirical motivation.

From a computational point of view, an expectation inspired by the theory of formal languages is that shifts within the computational morphospace have to do with the memory regime with which CSs operate. Such an expectation has been justified in the preceding pages both from the point of view of development and anatomy. From a developmental point of view, these perturbations have mostly extended the time—and rates of activity—of certain stages of embryonic and post-embryonic brain development; from an anatomical point of view, they have translated into a significant increase of the cortical/memory component of CS, with the basal/sequencer component remaining in contrast rather stable. Although at first sight this could seem a case of "mosaic" evolution, we claim that both facts are jointly derivable from the developmental dynamics of CS as an integrated system. Besides, the conservativeness of the basal component acts, within this model, as a force constraining the growth of the cortex, which probably means that the size of the latter in CS_{HUMAN}—as well as its associated functionality—is at, or very near to a limit point (Hofman 2001), which perhaps explains its apparently exceptional character. This does not mean, by the way, that size is the only relevant parameter explaining the computational power evolutionarily attained by CS_{HUMAN}; but it is probably the basic one, because other properties increasing the power and flexibility of computations—lamination, parcellation, and so on—can be explained as side-effects of cortical size—see Bullock and Horridge (1965), Ebbesson (1980), Deacon (1990a, 1990b, 2000), Strausfeld et al. (2000), and Striedter (2005).

We think that by putting all these pieces together a very coherent picture of the evolution of a crucial aspect of human cognition emerges. The next chapter is devoted to showing how this model can be extended to clarify the evolutionary connection between the human mind and the mind of other species, as well as to understanding the evolution of CCC_{HUMAN} beyond its computational underpinnings.

7

Other Minds

> Nevertheless the thought may be similar in man and beast. For when we assert that a man is running, our thought is no different from the thought that a dog has when he sees his master running.
>
> Thomas Hobbes, 1641

In this chapter we return to the question of the putative uniqueness of the Faculty of Language (FL) and, what it is more relevant, its evolutionary import. It is our opinion that this is a subject that deserves a more meticulous consideration than it has hitherto received. From a purely intuitive—and inevitably superficial—point of view, it seems indisputable that FL is a unique feature of humans. After all, no other species seems to speak the way we do and no other species seems to fully understand us if we direct our speech to it. However, a biologically informed approach to the issue requires a more carefully nuanced approximation. Generally speaking, in order to establish the degree to which a certain biological feature can be dubbed "unique," two fundamental questions must be addressed:

1. The component parts of the feature under consideration; and
2. Its developmental underpinnings.

The first question clearly suggests that queries on uniqueness are rarely—if ever—to receive all-or-nothing sorts of answers; the second question suggests—in agreement with the biological concept of homology (Wagner 1989a, 1989b)—that in order to resolve whether a certain feature is or is not species specific, factors and constrictions acting on development are more basic than their observable outcomes in the feature composition of adults. Besides, questions (1) and (2) are obviously intertwined, as anyone can clearly conclude from Minelli's (2003: Ch. 10) insightful comments concerning the main challenges of guessing homological relations from developmental grounds: What looks like a single or unitary feature when adopting a comparative perspective based on shallow observations can turn into a rather complex array of components when we descend to a developmental level of analysis, because the parts of organisms are mostly constructed by a factorial—or combinatorial—technique of sorts.

Back to FL, the case for uniqueness has been vigorously reopened in the last few years, starting with Hauser et al.'s (2002) take on the matter. A clearly positive note of this approach is that it goes beyond previous stances based on all-or-nothing bearings—clearly the case of Chomsky (1968) and related works, or Bickerton (1990); the negative counterpart is that it ignores the developmental perspective as the primary source of sound comparative information. It is our opinion that both the methodological proposal and the substantial tenets of Hauser et al. (2002) are seriously flawed by this shortcoming, in ways that we carefully comment on in the first section of this chapter.

The second section of the chapter is devoted to fully justifying our own contention—advanced at different points of the preceding chapters—that FL is connected in non-problematic—even if deep and subtle—ways with the feature constitution of any other species showing a complex nervous system—the minimal archetypical characters of which can be said to be present in vertebrates at large (Striedter 2005). Complementarily, if this developmentally founded comparative approach is on the right track, it offers us the opportunity of reaching a clearer understanding of the ways in which the minds of different species can evolutionarily depart from each other. Two basic developmental trends leading to inter-specific variation in patterns of mind/brain organization are defended in this section, of which we offer some concrete illustrations. This is an assumedly rather speculative section. Nevertheless, we think that it is an extremely critical part of this book, in that it explicitly points to the possibility of subjecting our evolutionary hypothesis to empirical tests.

A final section of this chapter is an application of our model of evolution as a possible tool for measuring the distance between the mind of modern humans and that of their closest known relatives, i.e., Neanderthals. This is the subject matter of an old and—we think—stagnant debate, to which newly available kinds of data have only served to add additional doses of confusion in the absence of a theoretical framework capable of setting out their real significance. We defend the argument in this section that our computational Evo Devo model can aptly serve to fulfill this role.

7.1 On the Many Senses of FL: Do We Really Need Them?

Hauser et al.'s (2002) position is, in a few words, that FL is and is not a unique feature of humans: It depends on how much of FL we decide to focus on when discussing the uniqueness issue. A more elaborated presentation of their thesis—and a less evasive looking one—is that FL is a complex organic system, which comprises (1) an innovative core computational component, and (2) a collection of rather conservative peripheral components, relating to (a) the ex/internalization and (b) the intentional content of the expressions generated by (1). They refer to (1) as "FL in the narrow sense"—FLN—and to (1) and (2) together as "FL in the broad sense"—FLB. FLB, on the one hand, is evolutionarily continuous with respect to the sensorimotor systems and the systems of thought subserving other activities in the case of other species; FLN, on the other hand, is an evolutionary novelty, which exhibits properties unknown in any other organic system, be it human or non-human. The property establishing the point from which FLN departs from anything known in nature—the story goes—is "recursion," by which these researchers specifically mean the capacity for unlimited structural embedding—see Fitch et al. (2005) and Fitch (2010) for some clarifications. Other properties of FLN such as long-distance and crossing dependences—of a higher level of complexity according to the theory of formal languages, are also of evolutionary interest, but their explanation presupposes the prior understanding of the exclusivity of recursion—see Chapter 5. Thus, recursion is the key evolutionary innovation brought about by FL—i.e., the property signaling the real gap between FL and any other organic system, and, consequently, the piece on which efforts must concentrate in order to solve the puzzle of language evolution.

So far, we have the substantial part of Hauser et al.'s (2002) contentions; another, no less important, aspect of their paper is that they invite us to think about it as a methodological recipe—i.e., as a particular way of dissecting the problem of the evolution of FL and a collection of clues concerning how to empirically test specific proposals on this matter. So understood, the main claim of their paper is that evolutionary explanations on FL build upon one or another for the following root positions:

Root positions on evolutionary linguistics

1. FLN is not human specific;
2. FLN is human, but not language specific;
3. FLN is both human and language specific.

Basic modus operandi

A human capacity other than language showing recursion refutes (3); a non-human capacity showing recursion refutes (2).

It is within this paradigm that some investigations have been devoted to show that recursion is a pervasive feature of human cognition—not just of FL—or that it can also be discovered in the minds of other species—not even closely related to humans in an evolutionary sense. Corballis (2007, 2011), who claims that recursion is a prominent feature of mathematical and social intelligence, is a clear illustration of the first tenet; Gentner et al. (2006), who contend that some avian species are sensitive to recursion when discriminating patterns of birdsong, is a good example of the second one.[1] Even if contradicting Hauser et al.'s (2002) substantial theses,

---

[1] It is worth commenting that Gentner et al. (2006) point to the possibility that recursion may be "cryptically" present in many species, because they detect it in starlings (i) after training, and (ii) in a non-spontaneous task—recognition of artificially compounded sequences.

researches like these are fully congenial with their general framing of the evolutionary study of language.[2] However, for reasons that will become clear in the following paragraphs, we think that they rely on a biologically ill-founded evolutionary framework.

To start with, note that Hauser et al.'s (2002) framework is based on the searching of an abstract property—"recursion," no matter what organic system actually exhibits it. We believe that this is an unjustified departure from standard practice in evolutionary biology, where discussions concerning evolutionary novelties are firmly grounded in the concept of "homology," which is a relation between structures, not between properties—i.e., abstracting away the structures showing them. So, when dealing with an apparently new structure, one reasonable first move in the application of the homological method is trying to break down the structure into its putative component parts, given the typically combinatorial—or factorial—nature of organs (Minelli 2003). This is an aspect of Hauser et al.'s (2002) framework regarding FL with which we agree. However, once a core computational system as a component of special evolutionary interest has been identified, we believe that it is a wrong next move to try to find out if some particular property of this system is also present in one or another system, regardless of whether the system in question can be reasonably deemed a true homologue of the Computational System (CS) of FL. Another crucial observation in this respect is that, in the end, true homologies are to be resolved attending to the developmental underpinnings of the corresponding systems, and not considering some partial similarities at a purely observational level.

Concerning all these questions, it is worth remembering now some tenets that we defended in the previous chapters and seeing how they fit—and if so—within Hauser et al.'s (2002) framework:

1. We claim, in agreement with them, that FL contains a CS as its core component;
2. We also claim, but now departing from their assumptions, that this system subserves not just FL, but a wide array of human cognitive faculties—hence, the adoption of the human Computational System (CS_{HUMAN}) as a shortened form for referring to it;
3. Also departing from Hauser et al. (2002), we further contend that CS_{HUMAN} is just a member of an extended family of homologous systems of computation, each particular system subserving very disparate faculties in different species;
4. Finally, we defend the contention that CS_{HUMAN} displays a high level of computational complexity, well above that of most experimentally tested and

---

[2] This line of experimental research starts with Fitch and Hauser (2004), where the contention is made that recursion is alien to the mind of at least some primates. We come back to this experimental paradigm in our conclusions for this chapter.

formally described CSs—CompuT1, but presumably also exhibited by other hitherto unattended systems—see below.

Let us concentrate on claims (3) and (4), the two most directly related to the current discussion. Contention (3), on the one hand, amounts to saying that all cognitive apparatuses that we identify as CS share a common ground of morphogenetic resources and constraints—which is, by the way, how this contention becomes open to empirical refutation; contention (4), on the other hand, means that variation within the space delimited by these morphogenetic parameters is discontinuous, an idea that opens the opportunity of establishing further subclasses within the general class of CSs—i.e., homologues in a hierarchically lower level of analysis. Taking all this into account, does it make any sense to maintain the FLN/FLB nomenclature of Hauser et al. (2002)? We argue in the rest of this chapter that the distinction is empirically dubious; besides, we believe that it is also to be avoided for conceptual reasons.

Forget for a while that we are to defend the claim that there exist systems of computation other than CS_{HUMAN} exhibiting CompuT1. Let us assume instead, for the sake of argument, that FL is unique in showing this particular level of computational complexity—i.e., that it is phenotypically discontinuous from any other CS. From this, one could conclude that FLN is a category that fits CS_{HUMAN} well. However, as we explained in the previous chapter, discontinuities at this level are not incompatible—quite the contrary—with continuity at the level of development, which—as we have repeatedly pointed out—is the essential one for evolutionary concerns. As a matter of fact, locating phenotypic discontinuities within the same morphospace amounts to saying that they are continuous from the point of view of development. From this, one should be inclined to conclude that FLN is not a suitable category for CS_{HUMAN}. So, CS_{HUMAN} is and simultaneously is not a biological discontinuity—i.e., it is and it is not FLN in Hauser et al.'s (2002) evolutionary sense, depending on the adoption of a phenotypic or a developmental perspective. Puzzlement does not stop here, however. For one thing: One could maintain—given the above-mentioned provisos—that CS_{HUMAN} is the only member of the natural class CompuT1—therefore, with no homologues, while still maintaining that it is just one member of the natural class CS—of which many homologue instances can be identified. So, again, CS_{HUMAN} is and simultaneously is not FLN in Hauser et al.'s (2002) evolutionary sense, depending now on the hierarchical level of homology analysis that is adopted.

Considering all this, we conclude that the FLN/FLB distinction is not worth preserving in that it is not flexible enough to deal with the subtleties of evolutionary analyses: Namely, it becomes unsatisfactory as soon as we incorporate developmental concerns into evolutionary explanations, a move that we consider mandatory especially when unearthing homology relations. So, instead of keeping this nomenclature alive at the cost of bewilderment, we suggest the following set of ideas and associated conventions:

1. FL is a cognitive faculty encompassing a core CS and a collection of sensorimotor and conceptual-intentional peripheral systems that the former system interfaces with—basically as in Chomsky (1995) and related works. This is the only concept of FL that is needed for evolutionary concerns.
2. An independent—but related—question is whether CS_{HUMAN} also interfaces with other non-linguistically related systems. In case it does—which is the position that we defend in this book, this would directly explain that many crucial properties of FL—starting with recursion—are also exhibited by other human cognitive capacities, such as social intelligence or mathematical thinking—in case one decides not to include them within the peripheral systems. This is a somewhat arbitrary decision, with no important empirical consequences. The theoretically relevant point is that—were this stipulation proven correct, it makes no great sense to consider properties such as recursion as subsidiary to FL in domains also manifesting them; they would instead be properties of a single system of computation subserving different domains of application—a departure, we think, from Chomsky's (1995) model. From an evolutionary point of view, it may be contended that they are "linguistic properties" inasmuch as CS_{HUMAN} can reasonably be said to be an evolutionary outcome "for" language. However, as we also believe—in agreement with Fodor and Piattelli-Palmarini (2010)—that the very idea of "evolved for" is conceptually out of kilter, we conclude that the question is theoretically meaningless.
3. Another relevant question is whether CS_{HUMAN} is an evolutionary novelty. We argue below that it is not, but this does not lessen its evolutionary interest. We contend that CS_{HUMAN} belongs to CompuT1, a natural class comprising—as far as we know—only a few members. A wider coverage of the question than has hitherto been done may eventually reveal that it is not such a rare family of systems. But even in this case the evolutionary interest of its particular instances will remain intact. In agreement with the model of "Morphological Evolution" presented in the previous chapter, they are not simply a matter of "descent," as they can be present in organisms belonging to distantly related taxa—a hint of which will be given in a while. It is thus of the greatest theoretical interest to try to settle the conditions leading to this outcome in very different evolutionary scenarios.
4. As for terminology, we will be using (i) CS to name the natural class of organs of computation—and also to refer to tokens of this class; (ii) CompuTn to refer to natural subclasses of such organs; and (iii) CS_{HUMAN} to refer to the particular implementation of CS—and also of CompuT1—subserving FL, among other

human faculties. We believe that this terminology is free from most of the shortcomings of Hauser et al.'s (2002) nomenclature.

In the following sections we put all these ideas and terminology into action to describe some paths of diversification of the computational mind, in a direct application of the methodological principles for the analysis of behavior we sketched at the end of Chapter 5. In the concluding section of this chapter we will compare our proposal with other, apparently analogous ones, where formal language and automata theories are also used as analytical tools in similar contexts.

7.2 On the Many Ways of Being a Mind

There are many different ways of being a different mind. One way your mind can be different from other minds is by having a particular type of CS, instead of another, alternative type; another way is by having a CS connected to a particular set of peripheral systems, instead of establishing an alternative pattern of connections. This view opens a high array of diversification in the architectural organization of minds, while keeping them essentially uniform—obviously, starting at a certain level of complexity of the nervous system. From this point of view, the expectation is that a linguistic mind is not very different from other, non-linguistic, types of mind. This is exactly the idea we are going to entertain in this section. Our aim is to explain and exemplify that FL does not presuppose a radically different mind/brain architecture. On the contrary, we will show that its composition simply consists of materials and patterns of connections also present in the mind of organisms in which nothing resembling language seems to exist. We interpret this fact as an illustration of the factorial—or combinatorial—character of FL and other similarly constructed mental systems, the explanation of which must ultimately be developmentally traced (Minelli 2003).

This rapid sketch of our ideas purports two main different claims: (1) That FL departs from other—non-human—cognitive systems in having a particularly complex CS, while sharing a similar pattern of external connections; and (2) that FL shares with certain cognitive systems a common type of CS, while departing from them in the corresponding patterns of external connections. A third logical possibility obviously exists: That a similarly complex CS associated to a similar pattern of peripheral connections also exists in some other species. But while this is an extremely interesting possibility, to the best of our knowledge it seems to have no empirical support—which, incidentally, maybe explains the feeling that humans are alone in having something remotely resembling language. In any event, later on in this chapter we carefully examine what we consider to be the nearest actual approximation to such a possibility—the mind of Neanderthals, our closest evolutionary relatives.

7.2.1 What is it like to be a Campbell's monkey?

Campbell's monkeys (*Cercopithecus campbelli*) have recently been the focus of considerable attention after the publication of a paper by Ouattara et al. (2009) demonstrating the variety and complexity of the vocalizations that they use as alarm calls, clearly exceeding that of previously known systems in related species—see Struhsaker (1967) and Cheney and Seyfarth (1990). A common feature of this family's (Cercopithecidae) alarm systems is that each signal refers to a particular type of danger—e.g., eagles, leopards, snakes, and so on, which in turn is associated to a stereotyped answer in order to avoid it—e.g., hiding in a bush, climbing a tree, surrounding and threatening the danger, and so on. In most cases, the inventory of signals seems to be very limited and each particular signal to be used only in isolation. Notwithstanding, a certain array of variation is known to exist concerning both parameters, with the system of Campbell's monkeys outranging that of any other species: On the one hand, the system categorizes different general types of danger ("predatory" vs "non-predatory") and then establishes further and more nuanced distinctions within each general category (e.g., "loss of eye contact with the rest of the group," "sudden fall of a tree or branch," "individuals from other groups," and so on, within the second one); on the other hand, most calls are not isolated signals, but combinations thereof. Let us observe with some details how are they organized.

The inventory of call units of this species comprises a total amount of six items—B, K, H, W_+, K_+, H_+ (the "+" symbol represents that the unit has an extra duration in relation to other units), some of which are exclusively used within combinations—i.e., they have no particular meaning in isolation. In all cases it is observed that, when combined with other items, the meaning of each particular unit is radically altered or completely canceled.[3] This observation implies that the semantics of this system does not obey the "principle of compositionality" which rules human language, according to which the meaning of the component parts of an expression is preserved in the meaning of the whole. In the case of the alarm calls of Campbell's monkeys, what we rather observe is that the signal as a whole is directly associated with a meaning that has little—if anything—to do with the meaning that the component units have—if they have one—when used in isolation. It is thus very clear that the "semantic design" of this system is quite different from that of human language, a question to which we will return later on. For the time being, let us concentrate on its raw computational properties.

Focusing our attention on vocalizations with a non-predatory meaning, we can observe, for example, that one repetition of B (BB) correlates with the meaning "loss of eye contact with the rest of the group," but if the same sequence (BB) is followed by

---

[3] Similar effects are observed in the relatively simpler vocalizations of other species of Cercopithecidae described in Arnold and Zuberbühler (2006a, 2006b).

a certain number of repetitions of K_+ ($BBK_+K_+\ldots$), then the new sequence correlates with the meaning "sudden fall of a tree or branch;" or, if the original BB sequence is followed instead by repetitions of H_+ followed by repetitions of K_+ ($BBH_+H_+\ldots K_+K_+\ldots$), then the new sequence is normally used as an answer to calls coming from individuals of neighboring or strange groups. Strange as this set of sequences may seem from an FL-centered perspective, the curious thing about it is that the formal complexity of the underlying CS can be fully captured by a single finite-state automaton—see figure 5.1, with each particular meaning corresponding to one of three alternative transition states within this virtual machine, as captured in figure 7.1.

Note that this figure represents the motif "BB" as a mandatory step for this whole family of "non-predatory" calls. From it, a probabilistic transition opens at which the composition of the signal can be predicted to follow one or another of three alternative paths: (1) Directly to the endpoint of the flowchart—"loss of eye contact with the rest of the group;" (2) to K_+ and, after one or more repetitions, to the endpoint—"sudden fall of a tree or branch;" and (3), to H_+ and, after one or more repetitions of H_+ and K_+, to the endpoint—"answer to calls coming from individuals of neighboring or strange groups." It is thus clear that, abstracting away the syntactic restrictions of the corresponding sequences, the CS underlying the alarm system of Campbell's monkeys and that underlying birdsong in Bengalese finches—described in Chapter 5, belong to the same computational type: Namely, CompuT3—i.e., a regular system with computational power equivalent to that of a finite-state automaton.

So, sophisticated as the alarm calls of Campbell's monkeys certainly are, they appear to lie far from the level of computational complexity of linguistic expressions. Units in each sequence are organized as "beads on a string," with each particular

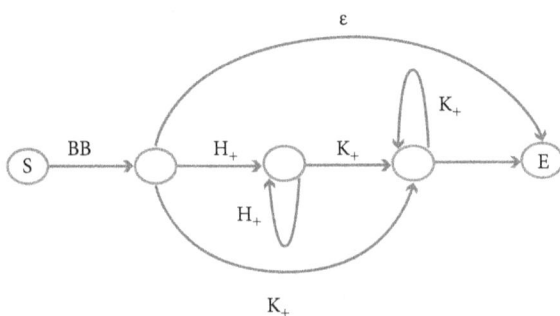

FIGURE 7.1 "This monkey's gone to heaven."
A nondeterministic finite-state automaton modeling the set of possible calls in the repertoire of a Campbell's monkey.
Caption from the lyrics of the song "Monkey Gone to Heaven" by the Pixies (1989).

element always being predictable from the preceding element.[4] This strictly linear style of arrangement strongly contrasts with that of human sentences, in which nested embedding—sets of units within sets of units—and cross dependencies—units related across those sets—are also pervasive features. As discussed in Chapter 6, this is not an obstacle to deeming the underlying CSs homologues "as systems of computation," as we strongly believe to be the case; but it means, at the same time, that they belong to qualitatively different computational phenotypes—CompuT3 and CompuT1, respectively, corresponding to different evolutionary trajectories within a common developmental morphospace.

But, in spite of this computational gap, we cannot refrain from feeling that the alarm system of Campbell's monkeys—and even the less sophisticated variants of other related species—is more "language-like" than other animal practices. This is an intuition that is not difficult to explain: When screaming, monkeys transform an otherwise internal representation into a perceptible input, which corresponds with something that we humans sometimes—*sometimes*—do when we speak. In other words, these monkeys' calls are "contentful" in a way that birdsong—which apparently has no conceptual import (Marler 1998; Marler and Slabbekoorn 2004)—is not.[5] Our technical way of expressing this intuition is by saying that CS, in the case of Campbell's monkeys—and maybe so in some other related species (Arnold and Zuberbühler 2006a, 2006b), interfaces with both the sensorimotor system and a particular system of thought, thus serving as a bridge connecting internal representations with intentional import and motor instructions and percepts, as represented in figure 7.2.

There are, of course, many obvious differences between the calls of Campbell's monkeys and human language. As we explained before, the monkeys' system is, from a semantic point of view, "non-compositional"—i.e., units lack a more or less constant meaning, preserved across combinations. A possible way of formalizing this feature is by saying that Campbell's monkeys lack a proper lexicon, comprising a collection of steady or fixed "sound–meaning" atoms. Or, in other words, that the external systems—sensorimotor system and system of thought—are more decoupled than in the case of FL, by which we mean that, in the case of monkeys, "sound–meaning" blends only—or mostly—obtain via CS. On the sensorimotor system side,

---

[4] No vestiges of systems above this level of complexity are known to exist in the different forms of "minimal syntax" (Ujhelyi 1996) of primates so far studied in the wild—for example, organisms capable of discriminating sequences like $BBH_+H_+H_+K_+K_+K_+$ from sequences like $BBH_+H_+H_+K_+K_+K_+K_+$ or $BBH_+H_+H_+K_+$, which would at least require a $CS_{CompuT2}$. Some experimental evidence also invites us to conclude that it could be beyond the cognitive capabilities of non-human primates (Fitch and Hauser 2004; but see Perruchet and Rey 2005 for some critical comments). Anyway, the question deserves to be more deeply investigated.

[5] This is not to say that alarm systems similar to those of monkeys are completely unknown among birds. They exist, but apparently consist of isolated calls (Griesser 2008), so—as is also the case with some monkeys' calls—they seem to be disconnected from a conceptual system.

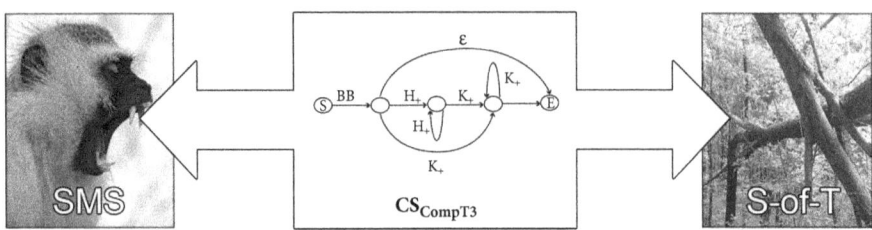

FIGURE 7.2 **Screaming trees**
The CS of Campbell's monkeys interfaces with a system of thought (S-of-T) and a sensorimotor system (SMS), making it possible for these animals to utter "contentful" calls.
The caption alludes to the name of the American rock band Screaming Trees.

an important difference is that articulatory gestures are more fine-grained and, above all, more subject to voluntary control in the case of language than in the case of the monkeys' calls. Anatomically, both features correlate with the fact that motor control is more corticalized in the former than in the latter, where it is mostly controlled by subcortical structures (Ploog 2002). According to our own hypothesis concerning the anatomical site of CS—see Chapter 5, the contention can be made that, in the case of monkeys, the sensorimotor system is less decoupled from CS than in the case of FL. Finally, on the system of thought side, a clear difference is that CS accesses a highly restricted array of categories, corresponding to very specific experiential domains—contrary to the domain-unspecificity of FL.

Different evolutionary hypotheses can be entertained concerning all these issues and the evolutionary outcome that we observe in the case of FL. It is possible, for example, that they are totally unrelated phenomena, each requiring independent evolutionary explanation. It is also possible, however, that they are—to a certain extent—different facets of a single evolutionary phenomenon. Suggestions in this latter direction can be found, for example, in Striedter (2005: 242–243), where the contention is made that the invasion of the spinal cord by neocortical axons—a side effect of the overdevelopment of the human brain, is probably responsible for the fine control of vocal gestures in speech. Besides, as modularization and inter-module connectivity are purportedly side effects of brain overdevelopment too,[6] the variety of conceptual domains simultaneously accessible to CS_{HUMAN} can also be reasonably related to the same evolutionary roots. These are however questions regarding which we cannot offer more detailed suggestions for the time being.

---

[6] See the conclusions of the previous chapter for some comments and relevant references.

7.2.2 What is it like to be a weaverbird?

This section—perhaps somewhat surprisingly, but crucially for our arguments here—deals with a specific form of construction behavior observed in a number of species of birds. It is important from the outset, however, to underline that our focus is not on construction behavior per se, but on a series of actions that some birds perform when building their nests. This is important because, as pointed out by Hansell (2000: 14), builders "do not need to be clever to be good at it, in fact it is not even necessary to have a nervous system."[7] As we will develop in detail in the last chapter, one of the interesting outcomes of our approach is that it brings into focus and offers a framework to investigate the problem of the origins of natural computational systems as one of the key innovations in the evolution of metazoan nervous systems. The implication here is that for the emergence of some behavioral pattern—even one that superficially may look very complex—it is not necessarily the case that an underlying CS exists, but that when there is evidence of its presence, we can assess its computational power on the basis of several factors including, of course, some observed behavioral pattern.

Turning then to birds' nests, it is again important to emphasize that not all nests are complex or the product of complex actions. Similarly, not all birds, even when they are capable of building apparently very complex and sophisticated structures, are good candidates for the attribution of a powerful CS,[8] since all building techniques identified by Hansell (2000, 2005), like piling up, molding, sticking together, or interlocking, do not appear to require very complex action patterns and are a clear illustration of the principle that "nest materials stay together both because of the spatial relationship established between them through the behavior of the builder, and because the materials have properties which hold them in that relationship" (Hansell 2000: 71). This principle, however, leaves out the two construction techniques we want to focus on here: weaving and knotting. These techniques are observed in two families within the order of the Passeriformes, namely the Ploceidae (or Old World weavers) and the Fringillidae—although, in the latter case, only in the Icterini, within the subfamily of the Emberizinae (New World weavers: Oropendolas, caciques, and orioles). Interestingly, as Hansell (2000: 81) points out, "in building technique these Old and New World weavers are markedly convergent;" despite the fact that their beaks are different in shape, it is the male who builds in the case of Old World weavers but the female in New World weavers, and they are very different in shape and size, with New World weavers having an average weight ten times heavier than that of most Ploceidae.

---

[7] Unless otherwise stated, our basic references for nest building and construction behavior in general are the two comprehensive monographs published by Mike Hansell in the early 2000s (Hansell 2000, 2005).

[8] At least from what can be deduced from their nesting behavior, which is not to exclude that complexity is observed in other areas; see below.

This convergent construction technique observed in weavers consists essentially of a variety of fastenings that the bird uses to attach strips of vegetation to one another in order to secure the nest in position. These fastenings "include loops, half-hitches, hitches, bindings, slip knots, and overhand knots, as well as a more regular over-and-under pattern of weft through warp that defines human weaving" (Hansell 2005: 73–74; also Collias and Collias 1962). Thus, as Hansell (2000: 80–81; see also his figure 4.13) points out:

> For weavers, the task is especially difficult; firstly, woven structures bear loads in tension and so the first strips must at least bear their own weight and, secondly, the strips have no inherent properties to secure them until tied to branches from which the nest will be suspended. Spiral wrapping round an attachment twig may give a strand temporary stability, but it must be secured with a hitch or knot; this may require integration of movements of beak and feet [...]

To emphasize this point, the task not only requires extremely fine motor control and coordination with the sense organs, but "[the behavior] also differs from that of foraging or grooming" (Hansell 2005: 74), to which it is often possible to relate many actions observed in the construction behavior of other birds and animals.

Finally, to complete this brief survey, even though the nesting behavior of weaverbirds appears to be instinctive—as is the case with all birds, in the sense that no explicit instruction or supervised learning appear to be involved, there is strong evidence that the quality of nests increases with the age of the individual and that lots of practice is needed in order to perfect the weaving technique. The clearest support for this conclusion comes from the experiment reported in Collias and Collias (1973), where young weavers were deprived of nest materials for variable periods of time resulting in "significantly retarded development of weaving ability" (Collias and Collias 1973: 371), which suggests "that some degree of learning enters into the refinement of each of the various steps leading to effective weaving" (Collias and Collias 1973: 381).

Curiously enough, even though Hansell (2005: 84–85) already observed that "tying shoelaces is a landmark in a child's development" and that consequently "weaving is the most difficult nest building technique for birds," little attention has been paid to it and most research on avian abilities has concentrated on other areas.[9] To our knowledge, the first to raise the question of the importance of knotting and weaving techniques for the assessment of cognitive capabilities were Camps and Uriagereka

---

[9] Most notably in birdsong and vocal learning as in the work by Okanoya (2002) reported in Chapter 5, but also in tool use (Hunt 1996; Hunt and Gray 2003; Lefebvre et al. 2002; Iwaniuk et al. 2009) or social behavior (Burish et al. 2004); see Emery (2006) for an overview of these and other aspects of avian cognitive abilities. Impressive as some of these abilities may seem, we are not sure that they are real indicators of the presence of very sophisticated computational capabilities and we remain agnostic as to their relevance, with perhaps one exception, bowers, which we will however leave out from our discussion, as they involve issues like "aesthetic judgment" that are difficult to assess in terms of computational complexity; but see Borgia (1986), Frith et al. (1996), Hansell (2000: Ch. 8), Madden (2001), and Hansell (2005: §5.5) for details and references.

(2006) in a paleoanthropological context—see next section, a work that later inspired our own first take on the issue in the field of avian cognition (Balari and Lorenzo 2009). The first point to be taken into account here is that, with the exception of weaverbirds, no other species apart from humans is known to tie knots in the wild,[10] which makes knot-tying a rare and almost exceptional ability in the natural world. But nature is full of rare and exceptional characters, such as echolocation in bats and electrolocation in gymnotoid and mormyrid fishes,[11] so what is it that makes knot-tying so special?

Camps and Uriagereka (2006) reason, albeit somewhat informally, that a knot cannot be the product of a simple Markovian process, that it rather can only be produced by a context-sensitive system. Should this contention prove true, we would have an indication that the cognitive resources necessary to produce a knot may be similar to those necessary to produce an expression in a natural language. There are several possible ways to assess the complexity of knot-tying. One, which we advocate here for reasons to be made clear presently, is to resort to the mathematical theory of knots. Knot theory is the branch of topology dealing with the nature and properties of specific kinds of mathematical objects known as "knots." Knots in topology are not really different from real-world knots, as they are conceptualized as elastic, closed, tangled strings, such that the most basic knot (the unknot) is just like a circle in space (i.e., a string joined by its two ends), but lying on a single plane—figure 7.3:

FIGURE 7.3 "My head's in knots."
(I). The unknot or trivial knot (left) and one of its many possible projections (right). Both entities are topologically equivalent, since they can be transformed into each other by a number of simple moves which do not involve introducing or undoing crossings in the string.

Both knot images were generated with the KnotPlot© software developed by Rob Scharein. Caption from the lyrics of the song "Knots" by Pete and the Pirates (2008).

---

[10] Some great apes in captivity have been reported to tie knots by Herzfeld and Lestel (2005); we'll come back to this presently.
[11] And in monotremes (Pettigrew 1999)!

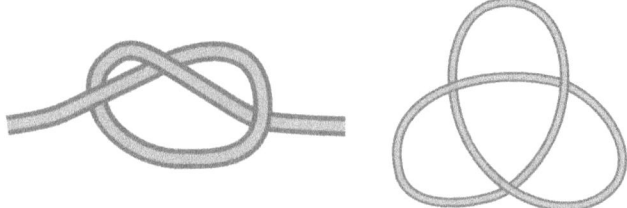

FIGURE 7.4 "My head's in knots."
(II). The trefoil knot—right—and the overhand knot—left. Topological knots are closed loops, unlike our everyday knots, which more closely resemble topological braids. Both knots and braids are equivalent, however, and, for example, both the trefoil knot and its corresponding braid are described by the same braid word σ_1^3.
The trefoil image was generated with the KnotPlot© software developed by Rob Scharein.

Other more complex knots can be constructed by producing a number of crossings of the string, so that some parts of the string now lie on more than one plane, with the minimal non-trivial knot being a string with three crossings, the so-called trefoil knot (figure 7.4, right). An alternative way of conceiving of a knot is in the form of a braid, where a braid in topology is a set of n strings attached to a horizontal bar at the top and at the bottom, such that each string intersects any horizontal plane between the two bars exactly once. The strings in a braid may be complexly tangled, forming woven structures, which are known to be equivalent to certain types of knots or, better, which are known to generate certain types of knots. Thus, a braid consisting of a single string with three crossings is equivalent to the trefoil knot and more closely resembles the familiar overhand knot (figure 7.4, left).[12]

Thus, all knots apart from the unknot are three-dimensional objects in 3-D space. An important area of knot theory is to determine whether a complexly tangled string is the unknot, and if it is not, what kind of knot it is. We will not enter into the mathematical details here, but depending on how the problem is formulated, it has been shown by Hass et al. (1999) that it is either in the **NP** class or in the **PSPACE** class in the hierarchy of complexity classes. Thus, Hass et al. (1999) show that the UNKNOTTING problem, consisting essentially in determining whether a tangled string is the unknot, is in the **NP** class. A generalization of this problem, the GENUS problem, is shown by Hass et al. (1999) to be in the **PSPACE** class; the "genus" of a knot is an indirect indication of the number of crossings of the knot, so that for example the genus of the unknot is 0, and, hence, the genus is also an indication of which knot it is. A later result, in this case due to Agol et al. (2002, 2005), reduces the complexity of GENUS to **NP**-completeness. Now, context-sensitive languages are

---

[12] For a general introduction to knot theory see Adams (2004), on which our presentation in the text is based.

known to be contained in the **PSPACE** class, but a number of mathematical results, consistent with the fact that natural language is only mildly context-sensitive, suggest that natural language processes have also the character of NP-complete problems (Ristad 1993). Thus the problem of determining whether a string is knotted or not is at least as hard as the problem of processing an expression in natural language, which might be taken as an indication that the computational power of CS_{HUMAN} is in fact at work when a human being performs one or the other task. Now the question is: Is the computational power of the Weaver's Computational System (CS_{WEAVER}) equivalent to that of CS_{HUMAN}, given the fact that both species are able to tie knots?

To be able to answer this question in the positive or negative further research is necessary, but there are a number of considerations that suggest that this might well be the case; in other words, and resorting to Owen's terminological distinction for types of homologies, that CS_{WEAVER} and CS_{HUMAN} are special homologues which maintain a relation of general homology with the computype $CompuT_1$.

Take first, the case of humans. We have been tying knots for tens of thousands of years but, crucially, we have also kept inventing new types of knots. This last point is important and is the one that justifies turning our attention to knot theory. Note that by conceiving knots as topological entities we are just focusing on their formal properties, without making any direct reference to the performance of the right motor sequences needed to actually produce the knot; we are concentrating on the mental capacity of conceiving of several possibilities of tangling a string in 3-D space, in other words, to be able to invent new knots, a variety of the UNKNOTTING problem, as we will argue below.[13] This is important because from the fact that one is able to learn or imitate some specific motor sequence we cannot infer the presence of complex cognitive capacities. Thus, so far, the scarce literature on knot-tying abilities in humans has only focused on how people learn to tie a knot either by instruction or by imitation (Michel and Harkins 1985; Tracy et al. 2003), but these studies tell us little about the human "faculty of knotting." Indeed, if we were to assess this putative "faculty" in the light of the results of these experiments we would probably deduce that it is certainly quite poor: For example, the knots used in Michel and Harkins' (1985) experiment are relatively simple (the sheepshank, the butterfly knot, and the "magic" slipknot), but only 37 percent of the subjects were able to learn to tie the three of them by just attending instructional demonstrations, i.e., by just observing the necessary motor sequence to tie them. Similarly, Herzfeld and Lestel (2005) report on an experiment in which some great apes were able to learn to tie knots. These apes

---

[13] To be sure, the process of inventing a new knot involves much more than just imagining a tangled string in 3-D space, since other considerations such as tension and friction are important. That's the reason why a good fastening does not necessarily always correspond to a topological knot. For example, the clove hitch is, topologically, the unknot but, given its formal properties, it is still an excellent way to secure some weight suspended on a line (Bayman 1977; Balari and Lorenzo 2010a). Thus the space of possible fastenings is larger than the space of knots, but this does not undermine our argument here.

(mostly orangutans, but also bonobos and chimpanzees) appear to have been trained in knot-tying or to have been able to observe knot-tying activities performed by their trainers (the paper is not entirely clear on this point), but all of them engaged in knot-tying activities if later provided with the appropriate materials. They nevertheless did only seem to be able to rehearse the knot they had learned to tie (essentially a shoelace) or to produce just hitches, which, topologically, are only varieties of the unknot. This suggests that these apes were merely able to learn a motor sequence by rote, but that they were unable to generalize over the act of knotting, although it would certainly be interesting to have additional data on the knotting abilities of great apes.

Be that as it may, this is, in our opinion, an indication that motor control plays only a secondary role in knot execution, invention, and learning, and that, rather, these three abilities involve at least a particular case of the more general problem of object recognition—i.e., knot recognition, and concomitantly that spatial representation abilities are also involved. This intuition is reinforced by the fact that, at least in the cultural traditions we know of, complex knots are taught not through the description of some hand gestures, but rather by resorting to mnemonic techniques whereby the learner is able to figure out *the number and the direction of the crossings* that make up the knot. For example, in our tradition, the bowline is often taught by telling the story of a tree (the longest end of the line) standing by a lake (a loop defined by crossing the shortest end over the longest one) out of which comes a snake (the short end of the line) that turns around the tree before going into the lake again, and which finds its counterpart in the following English rhyme: "Lay the bight to make a hole | Then under the back and around the pole | Over the top and thru the eye | Cinch it tight and let it lie."[14] Note that both provide the learner with spatial information about the number of crossings and their directionality.

Now, the literature on visual object recognition is abundant, but it is possible to identify an important trend where it is assumed that object recognition involves something akin to parsing in language. In the case of vision, a common view is to assume that object parsing involves the identification of a number of geometric primitives (often cylinders)—this is, for example, the approach of David Marr (1982) or of Irving Biederman (1987), to cite a couple of relevant examples. Underlying this is the assumption that spatial and object representation is entirely based on part-whole, or mereological, relations. This idea, however, has been subjected to several criticisms on the basis that parthood is insufficient to represent and recognize an object and that mereology needs to be complemented with the notion of connectedness, which is eminently topological. Thus, Casati and Varzi (1999), for example, contains a long and detailed argument in favor of the idea that object and spatial

---

[14] From <http://www.realknots.com/knots/sloops.htm>.

representation is mereotopological, not just mereological—a proposal that finds some support in certain experimental results suggesting that object recognition often does not involve parsing (Cave and Kosslyn 1993).

How does all this relate to (real) knots? If in spatial representation some topological relations like connectedness are critically involved, it may well then be the case that our claim that the topological theory of knots may be relevant is not that far fetched after all. Knots, real ones, have no obvious parts, just crossings in any of the three spatial dimensions, and form a connected whole, like mathematical knots or braids. Our contention is therefore that it is this information that is important at the time of producing a knot or figuring out one, in other words that to make a knot, one needs first to represent it and to represent it one needs to figure out its topology. This, we contend, is what is involved in knot recognition and, therefore, our appeal to the work of Hass et al. (1999) is, in our opinion, entirely justified, since they offer a proof that the UNKNOTTING problem is the class **NP** and this problem is a knot recognition problem. Further evidence in support of our speculations is, admittedly, wanting, but surely amenable to empirical verification both on psychological and mathematical grounds.

Let's go back to weavers. Unfortunately the literature on nest building by weaverbirds is scarce and sparse, and the most complete study still remains the one by Collias and Collias (1962), on which Hansell's (2000, 2005) surveys are also mostly based. Recall from our discussion above that weaverbirds, when building their nests, use a variety of techniques, including hitches, knots, and woven structures, but from the literature it is difficult to assess to what extent individual birds show some "creativity" in their work. Thus, Collias and Collias (1962: 574) point out that "[t]he unit movements used in nest building are quite stereotyped," clearly referring to the motor sequences followed by the bird while weaving. Soon after, however, they add:

All of the weaving is done with the bill, although the bird often uses one or both feet to help hold a strip. When a piece of nest material fails to stick on the initial push, the male often shifts to some other spot and tries again, and this exploratory pattern confers flexibility and adaptability to weaving.

The net result being that "[e]very nest is unique in the details of its fine pattern" (p. 576). This observation notwithstanding, the authors insist on the fact that "the repeated use of the same basic mechanisms in weaving [...] results in nests that look extremely similar to each other [...] and readily recognizable as belonging to the same species" (p. 576). Finally, on the basis of detailed observations of the different action patterns associated with the seven different stages of the building process, the authors conclude that nest construction is an entirely stereotyped and stimulus-driven behavior matching "the general conception of instinctive behavior adopted by many authors since the times of Darwin and Fabre" (p. 592).

We are not in a position to overtly challenge these conclusions, but we would like to point out that the mere observation of the action patterns of any construction behavior, of the sequence of its intermediate stages, and of the end product could easily drive one to similar judgments, even in the case of humans. For example, we invite the reader to consult descriptions of how pre-urban dwellings are built (e.g., those found in Schoenauer 1981) and to compare them with those of Collias and Collias (1962) for weaverbirds.

Much more illuminating is Collias and Collias's (1973) research on the development of nest building in weaverbirds. Jumping to their conclusions, we see that the development of a non-abnormal behavior depends on:

(1) practice (but not tuition) appropriately channelized by the growing structure of the nest itself, (2) improvement in ability to select and prepare nest materials, (3) maturation of the endocrine system, (4) integration of tendencies of both build and destroy nest, and (5) integration with a mate. (p. 380)

Focusing on the first two factors, notice that "practice" purports making play-nests and cooperative building among juveniles, and that total deprivation of building

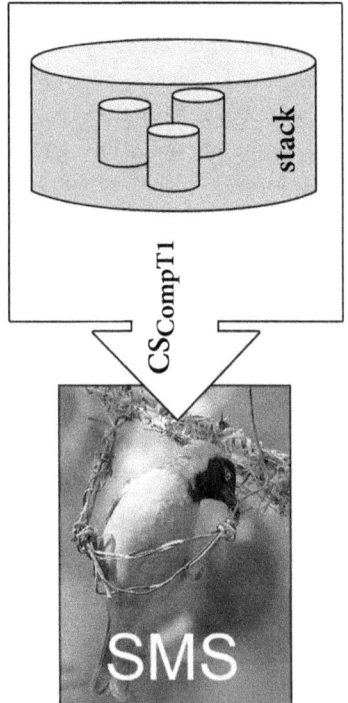

FIGURE 7.5 "Birdbrain"
Is the CS of weaverbirds homologous to that of humans, but interfacing instead with the spatial representation system and the sensorimotor system?
Caption inspired by title of the homonymous album by Buffalo Tom (1990).

materials for more than one year immediately after hatching results in the total incapability of building a nest, whereas partial deprivation (less than one year) results in severe retardation of the ability to weave and in the tendency to produce abnormal nests. This should moreover be compared to similar deprivation experiments with canaries reported by Hinde (1958), where no such effects were observed.

To conclude, and until further evidence is available, it may in fact be the case that CS_{WEAVER} and CS_{HUMAN} are homologous, although of course differing as to the components they interface with: The former being specialized in the development and control of knotting and weaving abilities (see figure 7.5), and the latter subserving linguistic and other abilities. Be that as it may, we believe that this example shows that the kind of research strategy we advocate here, where the focus is on the formal and computational properties of certain abilities and not on functional considerations concerning their use, may prove fruitful to evolutionary studies of language and other higher cognitive capacities.

7.2.3 What is it like to be a Neanderthal?

In this section we go immediately to what we consider the most compelling evidence so far pointing to the possession by Neanderthals (*Homo sapiens neanderthalensis*) of FL in exactly the same sense as anatomically modern humans (*Homo sapiens sapiens*). We refer to an intriguing collection of shells recently found in the cave of Aviones (Murcia, Spain), extremely interesting for different reasons (Zilhão et al. 2010)—figure 7.6:

1. They belong to non-edible species, which obviously means that they were collected for reasons other than nutrition;
2. They show remnants of pigments of different colors, indicative of an ornamental use and perhaps of symbolic connotations;[15]
3. They are perforated, so they could be used as beads;[16] and
4. They have been dated at approximately 50 ka[17]—i.e., before the arrival of modern humans in the Iberian Peninsula, thus excluding the possibility of interpreting the find as a case of acculturation.

---

[15] The oldest evidence of purportedly symbolic use of pigments is dated at near 300 ka in Barham (2002), where the contention is made that pigments were systematically and intentionally obtained by members of a transitional population of "archaic-to-modern" humans in Twin Rivers (Tanzania). The oldest attribution of a social use of pigments by Neanderthals is made in d'Errico and Soressi (2002), dated at 60–50 ka (Pech-de-l'Azé, France). McBrearty and Brooks (2000) observe that, anyway, the use of pigment by European populations at that and even later dates is extremely scarce if compared to its use in Africa at the time by anatomically modern humans.

[16] The oldest known marine shells perforated for ornamental use are those found in Blombos (South Africa), dated at 70 ka and associated to anatomically modern humans (d'Errico et al. 2005).

[17] Actually, between 45,150 (±650) and 38,150 BP (±350)—Zilhão et al. (2010: Supporting Information III).

FIGURE 7.6 **The Neanderthal's necklace?**
The evidence is inconclusive; see discussion in the main body of the text.
Original image from Figure 1 (detail) of Zilhão et al. (2010) and used with permission from the National Academy of Sciences, USA. The caption is inspired by the title of Juan Luis Arsuaga's (1999) book.

Why we consider this find so exciting is—we think—straightforwardly clear. First, if these shells were really used as beads then they should be threaded and the piece of string—whatever the material being used—somehow knotted.[18] So, this practice should be associated with spatial and motor skills connected to a CS of a type higher than CompuT2—see Camps and Uriagereka (2006), where using knots as a test for measuring the antiquity of FL was originally proposed, and Balari et al. (2011), as a more recent elaboration of the idea. Besides, if shells were pigmented for symbolic purposes—in which case most probably having to do with social status, these computationally sophisticated skills would also be connected to some systems of thought—at least to a module of social intelligence, were the previous comment somehow confirmed. In conclusion, underlying these putative Neanderthal necklaces there would exist a capacity of combining meaningful units associated to a CS within the T1 area of the Chomsky Hierarchy—i.e., a CompuT1.[19] We have no empirical basis to take the picture further—did combinations exhibit complex patterns? Were combinations, and not just units, also meaningful? Were combinations meaningful in a compositional way? Etc. However, as words and phrases do not fossilize, it is our opinion that the pigment-stained shells of Aviones are the closest evidence ever of an indirect proof of the presence of complex linguistic skills in a species other than anatomically modern humans. So they deserve to be meticulously examined and interpreted within the context of the recent explosion of information concerning Neanderthals.

To be fair, there is not a single, straightforward interpretation of these pieces. The conclusion that Neanderthals were capable of tying knots—against Camps and

---

[18] Knots are not directly attested to in anatomically modern humans until 27 ka B.P., by means of weaving, both in clothing and clothing representations—see Soffer et al. (2000).

[19] We hope that this explanation is enough to fulfill Botha's (2009) exigency of making explicit and convincing bridging arguments linking beads to language, which he directed against d'Errico et al. (2005).

Uriagereka's (2006) expectations—and that this ability was connected to the capacity of symbolic representation in certain domains favors the idea that they were also capable of computing complex strings in other domains. Their having an FL with a basic design not very different from that of modern humans then presents itself as a rather plausible hypothesis. Furthermore this scenario would rationalize Krause et al.'s (2007) finding concerning the antiquity of the human variant of *FOXP2*, a gene purportedly implied in the development of FL—see Chapter 5 and references therein. Contrary to previous estimates (Enard et al. 2002), Krause et al.'s sequencing of fossil strings of DNA from some Neanderthal individuals of more than 40 ka demonstrates that two point mutations previously considered specifically modern were already present within the Neanderthal genetic pool. So they conclude that whatever their developmental role in the case of modern humans, it was already operating in the development of Neanderthals. Krause et al.'s contention is that genetics thus seems to support the position that FL is a sapiens synapomorphy—and not a modern autapomorphy, already present in the last common ancestor of Neanderthals and anatomically modern humans some 300 or 400 ka ago.

Krause et al.'s argument, however, is not without some shortcomings. Leaving aside technical questions—but see Coop et al. (2008)—and even some commentators' skepticism over whether these specific mutations are actually relevant to language development in the case of modern humans—see Ptak et al. (2009), some caution is necessary before accepting Krause et al.'s conclusion, considering that *FOXP2* is a regulatory gene whose activity relates to hundreds of other genes. This simple observation has far-reaching consequences, because it means that the said mutations could have very different effects in different molecular contexts. The fact that in the case of humans it has an impact on the regulation of language development is actually a function of both the regulatory gene itself and the system of regulated or target genes—see Spiteri et al. (2007) for some information regarding this system in the case of humans. So, some information concerning the putative Neanderthal regulatory system would also be required in order to settle the question by just genetic means—see Benítez-Burraco et al. (2008) and Balari et al. (2011), for more detailed comments on the matter. No such information exists so far.

So we think that Krause et al.'s argument is in itself inconclusive as a proof of the existence of a Neanderthal version of FL—$FL_{NEANDER}$. Arguably, the Aviones find may be interpreted as independently proving that *FOXP2* was already acting on the development of a $CompuT_1$ within Neanderthals, exactly as in the case of modern humans. But again, given the inconclusiveness of Krause et al.'s contention, one would expect particularly strong support of the claim that the shells from Aviones were actually used as beads. However, no such support exists. We know for sure that anatomically modern humans intentionally perforated the shells from Blombos Cave some 70 ka ago (d'Errico et al. 2005);[20] however, the shells from Aviones were indisputably perforated by natural causes (Zilhão et al. 2010). Obviously enough, this fact does not exclude the possibility that they were collected with the purpose of using them as beads—as Zilhaõ et al. contend. However, it seems rather implausible that individuals capable of threading and knotting had relied on accidental finds for the collection of these appreciated items, given that the techniques required to puncture them, as well as the associated visuo-manual and computational requisites, are obviously much more elemental. Note that we are not arguing against the ornamental and even the symbolic uses of these shells. For us, points 1 and 2 above are indisputable; it is just point 3 that we believe needs further empirical support.

In any event, let us suppose for a while that the question was somehow settled and that we can also be confident of point 3. Within this scenario, it becomes very intriguing to try and make sense of the fact that Neanderthals had a CompuT1 while at the same time exhibiting such a different behavioral repertoire compared to that of modern humans—see d'Errico et al. (2003) and Balari et al. (2011). In domains of experience more closely connected to the symbolic aspect of FL, it seems that Neanderthals did not make use of any form of notational system (d'Errico et al. 2003), the practice of which became very common at the end of the Upper Paleolithic.[21] The practice of intentional burials is already well documented among Neanderthals in the late Middle Paleolithic—more than 100 ka, according to Grün and Stringer (2000). However, while burials of modern humans usually contain offerings—implying rituals with a symbolic import, Neanderthals' burials lack this symbolic component (d'Errico 2003)—so they are better interpreted as a hygienic practice (Mithen 1996; Klein and Edgar 2002),[22] or as the expression of social or emotional links (Mellars 1996) or the importance of ancestors (Mithen 1996). Besides, there exists no evidence of musical traditions—hints of which are dated at 35 ka in the case of modern humans (Conard et al. 2009), or other forms of art—which start very early in the Later Stone Age in Africa in the case of modern humans (Henshilwood et al. 2002; Lewis-Williams 2002) and experience a true explosion after their arrival in Europe (Pfeiffer 1982; Mellars and Stringer 1989).[23] Again, there does not exist a single, straightforward interpretation for these contrasts.

Some authors favor the idea that Neanderthals did not behave like modern humans simply because they lived under very different environmental and

[20] More ancient perforated shells have been found (Bouzouggar et al. 2007; Vanhaeren et al. 2006), but the nature of the perforations is disputed.
[21] d'Errico et al. (2003: 31) make the suggestion that the geometrical patterns engraved on pieces of ochre found at Blombos Cave are perhaps the earliest form of such a system of external memory.
[22] This practice is not unknown among non-human species—see Pinel et al. (2003).
[23] An exception to this statement seems to be the Châtelperronian and other late Neanderthal local cultures in Europe. It is a debatable question whether they were independent cultural developments (d'Errico 2003) or, rather, the product of emulation of the recently arrived modern populations (Coolidge and Wynn 2004; Mellars 2005)—a position we believe much more likely.

populational pressures, which made many of the moderns' achievements unnecessary—i.e., these achievements were within the reach of Neanderthal cognition, but Neanderthals did not need them (Finlayson 2009). The fact that, after all, there does exist some overlapping between the symbolic repertoires of Neanderthals and modern humans can be interpreted in this direction. For other authors, however, there existed a true biological gap between these species—which made it impossible for Neanderthals to share the worldview and practices of anatomically modern humans (Klein 2009). Within this perspective, one can tentatively conclude that Neanderthals—as well as their most immediate ancestors—had already evolved the component parts of FL, but that their connections were not well established and fully operative—the reflex of which are the asymmetries commented on so far.[24] Furthermore, such a conclusion is compatible with several recent studies that point to the existence of subtle differences in the ontogenetic paths of Neanderthals and anatomically modern humans, which—even if subtle—could maybe have a far-reaching impact on the respective patterns of brain organization—see Coqueugniot and Hublin (2007), Ponce de León et al. (2008), Gunz et al. (2010), and Gunz et al. (2012). We believe that this is a very likely scenario, taking into account that many side effects of brain growth within a model of concerted evolution—see Chapter 6—are very seemingly sensitive to the particulars of developmental paths. So a minimally prolonged and faster pattern of brain growth might have a large impact in parameters such as parcellation (Ebbesson 1980) and cortical invasion (Deacon 1990a, 1990b, 2000)—and, correspondingly, offer a clearer definition of the component parts of FL external to CS and, crucially, cause an increase of inter-module connectivity, not an unlikely explanation of the cognitive/behavioral novelties brought about by modern humans–see Mithen (1996).

We are afraid, however, that this is a question that cannot be empirically settled for the time being. But, for us, what is at least clear is that by discussing whether the said gap existed—or not; or to what extent it existed, we are really discussing whether Neanderthals did possess—or not—a CompuT1 system with–or without—the relevant connections established to compound—or not—a true FL. This is a non-

---

[24] There exist, of course, alternative explanations to these asymmetries. For example, Wynn and Coolidge (2011: 7-9) relate some of them to the evolution of an enhanced system of working memory (EWM), which allowed the externalizing of information for the first time and its combination with internally ongoing mental processes. It is worth clarifying that "working memory" (WM) in Wynn and Coolidge's sense does not correspond with our own sense of the term. For them, WM is an attentional function, complexly articulated—essentially as in Baddeley (2007)— and accessible to conscious awareness; for us, it is a space in which certain operations hold at a subattentional and subconscious level, itself a component of a larger function—namely CS. However, Wynn and Coolidge's interpretation of the evolution of modern cognition is not a priori incompatible with our own interpretation of the above-commented asymmetries, as we think that one of the effects of CS_{HUMAN} was that of enhancing attention— see Balari and Lorenzo (2008: 302), instead of the other way around. See Klingberg (2009) and references therein, for a model of the relation between memory and attention more congenial to our vision.

negligible question, because it can be instrumentally useful in focusing our attention on new domains where the activity of such a system—or parts thereof—could be reasonably expected to operate. Previous approaches to the question, operating with intuitive and loosely articulated concepts of language, were far less suitable for offering such new avenues of research and for honoring the promise to invigorate the long-stagnant debate over the Neanderthals' cognitive and linguistic capabilities—see Balari et al. (2011).

7.3 Conclusions

In this chapter we have illustrated our proposed novel methodology for the analysis of behavior with three case studies: The calls of Campbell's monkeys and the purported knotting abilities of weaverbirds and Neanderthals. The basic analytical principle on which this methodology is based essentially derives from the study of the formal properties of some behavioral activity (e.g., monkeys' calls) or of some of its products (e.g., knots) in order to find some suitable mathematical model (e.g., finite-state languages or topological knots and braids) to which these can be reduced in order to assess their computational complexity. In some sense, then, our proposal can be seen as an extension of an experimental paradigm originally initiated by Marc Hauser and W. Tecumseh Fitch, based on the aural pattern-recognition abilities of cotton-top tamarins and reported in Fitch and Hauser (2004); see also O'Donnell et al. (2005). The same paradigm was later applied to starlings, with seemingly equal success by Gentner et al. (2006), although it has been subjected to several criticisms from different quarters (e.g., Perruchet and Rey 2005; van Heijningen et al. 2009; Petersson et al. 2010; Rogers and Pullum 2011). The essence of all these criticisms is that the ability/inability of some animal to learn some complex pattern presented aurally is inconclusive with respect to the ability/inability of the animal to learn/use rules of some specific complexity. The problem can be easily illustrated with a simple formal example. Consider, first, the string $s = aaabbb$. Now, it is obvious that the following three statements are all true:

(1) a $s \in L1 = a^*b^*$.

b $s \in L2 = a^n b^n$.

c $L1 \cap L2 = L2$.

Thus, s can be weakly generated by both L1—a regular language—and L2—a context-free language; (1c) is just a particular case of the more general mathematical result that the intersection of a regular language with a context-free language is a context-free language. Now, suppose we want to train some animal to identify strings following the pattern defined by L2 but excluding the pattern defined by L1. In this case, we would take a set of strings belonging to L2 and we would use them to train the animal to recognize them. Inevitably, our training stringset would be finite and the subsequent test-rounds to verify if the animal learned the pattern would also be finite. Now, the problem is that, from the statements in (1), it follows that a (quite large) deterministic finite-state automaton is capable of generating L3 = $\{a^n b^n \mid$ for some fixed value of $n\}$, a finite (but possibly fairly large) subset of L2, meaning that we could never be sure of whether the animal has really interiorized a set of rules of the desired complexity or it is just resorting to a strategy using simpler rules while still being capable of generating a finite but substantially large set of elements showing the desired pattern. There have been some attempts to fix this particular problem within the same experimental paradigm (e.g., Rogers and Hauser 2010) and it is not our task here to judge whether these have been successful or not. We would like to claim, however, that our methodology may be a useful complement to aural pattern-recognition experiments, since it proceeds in the reverse direction, from the behavioral patterns the subject is naturally capable of performing to their complexity, such that the results obtained could then be matched to those of some aural pattern-recognition experiment in order to verify if some coincidence is observed.

The relevant sections of this chapter have been named after a celebrated paper by Thomas Nagel—"What is it like to be a bat?" (Nagel 1974). Nagel's thesis in this piece is twofold. On the one hand, he argues that we—human beings—cannot imagine what it is like to have experiences like those of bats—or wasps, flounders, or any other kind of organism complex enough to attribute relatively rich experiences to. This is, according to Nagel, because in order to properly accomplish such an imaginative exercise, we should be structurally conformed like those organisms—or approximately so. On the other hand, he defends the view that things are actually more extreme than that, as no one can imagine what it is like to have other people's experiences. According to Nagel, experiences have a subjective character, and we humans have not yet invented a method to objectively observe and describe this kind of phenomenon. Curiously enough, the first problem—"the bat problem," seems much more dramatic to Nagel than the second one—"the other people problem." For him, while the latter could be approached by developing a suitable phenomenological method, the former is definitely beyond the reach of a humanly organized being.

Nagel is maybe right—but perhaps only partially. In this chapter we have shown that windows can be opened to glimpse into the mind of a Campbell's monkey, a weaverbird, or a Neanderthal—and, surely enough, a wasp, a flounder, and even a bat. This is not to say we can fully imagine the experiences of these organisms—after all, other organisms' experiences also have a subjective character. However, glimpses of their mental life can be obtained—and probably very vivid ones—concerning facets of their cognition underlying which we can reasonably expect the functioning of the same organ of computation—obviously, "under every variety of form and function." This observation would be of little use, were it not for the fact that—we think that against Nagel's expectations—homologues seem to be widespread at this level of biological analysis. In the end, organisms–even bats—are not so differently structured. This is not to say that we have found the method that will allow us to imagine what it is like to have the experiences of a monkey, a bird, an extinct human, or a bat—of course not. But it clearly means, however, that we now have a key to obtain useful representations of what it is like to be any one of these organisms. We think that this is a goal worth pursuing, both theoretically and empirically.

8

Conclusions

> So... even if... Bugs do it/ And molds do it/ It's only 'cause their sensory mani-/Folds do it. And that's not good enough.
>
> Jerry Fodor, 1986

> It is very often inadequate to frame questions in terms of origin and evolution of an organ or an organ system, such as the brain or the circulatory apparatus.
>
> Alessandro Minelli, 2009

Throughout this book we have been spinning a dense fabric of speculations, empirical evidences, theoretical arguments, and tentative proposals with the aid of such disparate disciplines as linguistics, computational science, developmental biology, philosophy, anthropology, ethology, and many more. The result we hope, is a framework within which interesting hypotheses can be framed on the phylogeny and the ontogeny of cognitive capacities in the broadest sense, i.e., not only about human language or human cognition. It is time to end our deliberations and to pause for reflection. To do this, we propose speculating a little further on some areas in which we believe our approach may prove fruitful.

As we pointed out in the previous chapter, what started out as a quest on the evolution of language suddenly turned to the evolution of computational systems, and we wouldn't like to take leave without briefly touching on this issue once again. Moreover, given our characterization of Computational Systems (CS) as the natural counterparts of digital computing machines and given the severe criticisms wielded against the "computational paradigm" since, at least, the late 1980s we would like to spend some time to make this specific claim more precise within this evolutionary context.

8.1 Do Cnidaria Have Mental Representations?—A Final Speculation on the Evolution of Computational Systems

Jerry Fodor once asked why paramecia do not have mental representations (Fodor 1986). Not surprisingly, he was assuming from the outset that paramecia lack something that we humans (and perhaps other creatures) have; more surprisingly perhaps, his answer was not "Because they don't have brains." Presumably, this would most closely resemble the answer John Searle would have offered to the same question (Searle 1980). No, Fodor's answer is more sophisticated and deserves some space in this concluding chapter, as it condenses a number of ideas we have developed in detail throughout this book—not without problems, however, as we shall see presently.

In essence, Fodor's paper was meant as a reply to a number of criticisms raised by Daniel Dennett (Dennett 1978: Ch.6) concerning the Language of Thought Hypothesis and the Representational Theory of Mind (Fodor 1975) as well as a counterattack to Dennett's own take on the issue of mental representation, namely the so-called Intentional Stance (Dennett 1978: Ch.1, 1987: Ch.1). One does not need to undertake a detailed scrutiny of Fodor's and Dennett's work to discover that their main source of disagreement is not computationalism, but the nature (and perhaps also the origin) of intentionality, as both agree on the idea that some version of computationalism or other will at some time explain at least an important part of the functioning of minds.[1] To be sure, for Fodor both issues are intimately connected, as he has always contended that the only way to explain behavior is to assume that the objects of computational operations are primitive bearers of content (see also Pylyshyn 1984), therefore, his answer to the paramecia case partly relies precisely on this assumption. Thus, so Fodor's story goes, the behavior of a paramecium (and of all paramecia-like creatures) can be predicted solely on the grounds of their ability to react to physical properties of stimuli. In other words, "there is a lawful connection between a certain property of the 'stimulus' [...] and a certain property of the ensuing behavioral response" (Fodor 1986: 9). Fodor calls these properties, like temperature, light

---

[1] We use the word "computationalism" roughly as synonymous with the assumption that the activities performed by natural CSs are Turing-computable, i.e., that they can be formally characterized by such devices as the automata of the Chomsky Hierarchy. Our use of the word here is therefore neutral as to the kind of computational architecture that is assumed. Actually, it is a use that encompasses even Dennett's position, for example, which is not easy to pin down, but nevertheless congenial with the truth of the Church-Turing Thesis (see Dennett 1987: Ch.3, 1991: Ch.7, 1995: Ch. 15). The validity of the Church-Turing Thesis has been called into question—and the entire program of cognitive science hence claimed to be hopeless—by for example Roger Penrose (1989, 1994) under the assumption that some form of super-Turing computation or hypercomputation is possible; see Copeland (2002) for an overview and Davis (2004) for a critical appraisal of hypercomputation. The fact is that even though some architectures have been claimed to be able to go beyond the limits imposed by the Church-Turing Thesis (Port and van Gelder 1995; van Gelder 1995, 1998) there is no evidence at all that such extra power is observed in the actions of natural computational systems like, for example, human FL. Thus even if hypercomputation turned out to be a possibility, we believe that our position (and Fodor's and Dennett's) would not be undermined by this as it would merely add the theoretical possibility of much more powerful systems than those so far envisaged to our hierarchy of computational phenotypes. The kinds of objections posed by the hypercomputation movement are thus orthogonal to the kinds of objections to "classical computationalism" (Scheutz 2002) presented by those who allegedly defend alternative views of cognitive architecture such as connectionists and dynamic system theorists; we will touch on this latter issue below.

intensity, and so forth, *nomic properties*, but he warns of the fact that not all properties of the objects in the world are nomic (such properties as "being big" or "being dangerous," for example) and that some creatures (e.g., humans) appear to be capable of selectively reacting to such *nonnomic properties*. For Fodor, the only way to explain the behavior of such creatures (i.e., to assume that it is also lawful) is to assume that through some psychophysical relation between some creature C and some object O, the nomic properties of O are used by C to represent O as having some nonnomic property in virtue of which C will behave in some way or another. In other words, the observed behavior can only be explained by "introducing a *semantic* connection into the causal chain" (Fodor 1986: 14; emphasis in the original). Thus, according to Fodor, there is a great divide in the phylogenetic continuum separating those creatures only capable of responding to nomic properties from those capable of also responding to nonnomic ones, namely those creatures capable of mental representation.[2]

Now, as we advanced above, we deem problematic Fodor's picture of the phylogenetic divide as it is probably not fine-grained enough to capture the enormous variation that certainly exists among metazoan CSs and it moreover invites a reading of metazoan phylogeny as a process in which the attainment of mental representation truly defines a landmark in the history of life. At the same time, it nonetheless contains the seed to elaborate some interesting hypotheses about how CSs could have evolved. The first thing to note in this connection is that the distinction between one and the other kinds of creatures may be directly connected to two of the main hypotheses that characterize Fodor's thought: (i) The hypothesis of the modularity of mind (Fodor 1983), and (ii) the hypothesis of the language of thought (Fodor 1975). The connection with (i) is the idea that action is mediated by the intervention of a number of specialized modules or "vertical faculties" whose output is later sent to the central systems responsible for planning behavior; the connection with (ii) is the assumption that for both modules and central systems to be able to perform their activities some kind of representational medium (or "language") is necessary. Fodor's answer to the paramecia question can therefore be recast in the following terms:

(a) Nonnomic Sensitive Creatures (NNSC) possess certain specialized modules that, through the action of their interface systems, are able to represent in some internal language certain objects of their environment as having some nonnomic property on the basis of the input furnished to them by sensory

---

[2] As an aside, note that perhaps Dennett would be happy to endorse Fodor's proposal as his instrumentalism doesn't appear to deny the existence of representations but rather the idea that there exists something like intrinsic (or original or primitive) intentionality—here is where Dennett parts company with Fodor, see Dennett (1987: Ch. 8). Another source of disagreement between them, of course, concerns Dennett's faith and Fodor's skepticism on adaptationist explanations. For our part, as already pointed out in Chapter 4, we pronounce ourselves agnostic about the nature of intentionality but, of course, we stand on Fodor's side in all that concerns adaptationist accounts of the evolution of cognition.

transducers. This information then enters the central processors to plan behavior.

(b) Nomic Sensitive Creatures (NSC) react directly to the output of sensory transducers without the intervention of any specialized module and, hence, of any representational language.[3]

Now note that our notion of a natural CS in fact corresponds to a proper part of a Fodorian module, namely to the core of its syntactic engine. From this perspective then one of the corollaries of the proposals we have developed in this book is that variation is expected along (at least) two different dimensions: (i) The computational power of the CS that some organism possesses, and (ii) the input systems it interfaces with. Thus for example, in the case of Campbell's monkeys discussed in Chapter 7, we could speak of an "alarm call module" made up of a CS with a power equivalent to that of a finite-state automaton, interfaced to a sensorimotor system and to a conceptual system of the appropriate kind; similarly in the case of weaverbirds we could speak of a "nest-building module" (or, perhaps, just a "knot-tying and weaving module") made up of a CS with the power of a context-sensitive machine, interfaced to a sensorimotor system and to, say, a "knot and braid representational system." Clearly, this analysis does not preclude the possibility that the said CSs interface with other input systems to make up other specialized modules nor that other observed "behaviors" in these organisms are not mediated by specialized modules at all and follow the nomic path. We believe the latter to be an important consequence as it adds the appropriate degree of granularity that the original idea by Fodor is lacking: The question is not one of having or not having representational power, but rather one about in which domains representations are used.

Indeed, in this slightly modified scenario the distinction between NSCs and NNSCs becomes less clear-cut than the original proposal by Fodor appears to suggest, as it entails that both modes of "processing" may actually be present in the very same organism.[4] Our tentative hypothesis concerning the architecture of "metazoan minds" could thus preliminarily be summarized in the following points:

---

[3] Strictly speaking, then, as Fodor (1986) would put it, NSCs do not "behave," but they just react or respond in a rigid way to some physical stimuli.
[4] One might argue at this point that our idea that the CS may be shared by different modules is a perversion of Fodor's original definition of module, but recall that Fodor himself was from the beginning ready to accept that some cognitive system might be modular "to some interesting extent" (1983: 37), meaning, for example, that sharing some "horizontal resources" like memory or attention with other modules would not necessarily undermine its modular nature. In fact, Fodor takes informational encapsulation to be the most important aspect of modularity and this in our account is ensured by the autonomy of the interface systems (i.e., the bodies of knowledge or Chomskyan modules) into which the CS is plugged. Another point before we proceed: Note that our elaboration of the modularity thesis is entirely congenial with Fodor's also in the sense that it just assumes that some faculties may be modular, not all of them. Crucially, it does not entail the Massive Modularity Thesis typically associated with the partisans of evolutionary psychology (Pinker 1997; Plotkin 1997) and which Fodor has always contended to be

(i) Some behaviors in an organism are subserved by Fodorian modules, while others are not.[5]
(ii) Fodorian modules have a "classical" architecture in the sense that they contain a CS that works as a syntactic engine that manipulates mental representations.
(iii) To have a "classical" architecture is however not necessarily a property of the whole mind of the organism, which may combine both classical and non-classical modes of processing.[6]

In essence, from an evolutionary point of view what this hypothesis entails is that certain organisms have come to develop computational modules with the properties just described to control specific behaviors. Since the hypothesis rests on the controversial assumption that the architecture of such modules is classical, we would like to meet the challenge and try to elaborate our proposal a little further.

Classical computationalism has come under attack from different quarters and perspectives. We have already mentioned the case of Penrose (1989, 1994), to which we could add those who claim that we do not yet understand what computation really is (Smith 2002, 2011), those who argue that even if computation is acceptable as an abstract notion the answer to the question of its realization (or implementation) in a physical system makes it inappropriate for cognitive theorizing (Putnam 1988; Searle 1992), and those who claim that other, alternative paradigms like connectionism or dynamic systems are much better suited to explain the functioning of minds (Churchland 1995; Smolensky 1988, 1995; Port and van Gelder 1995; van Gelder 1995, 1998). As neither Penrose nor Smith have offered a real alternative yet we will leave them out of our discussion; just like Putnam and Searle, whose arguments appear to be not as damaging as it may seem (Scheutz 1999). Connectionism and dynamic systems theory however, might really pose a challenge to the idea that some cognitive activities are to be modeled as the performance of digital serial machines that manipulate symbols; moreover, in Chapter 6 we invoked dynamic systems theory in our account of developmental systems, so we deem it important to clarify what at first may seem a contradiction in the embracing of two apparently incompatible frameworks.[7]

---

incongruous (Fodor 1983, 2000). Anyway, we must confess that our commitment to the modularity thesis is weak, and we wouldn't be against weakening it even further if, for example, Ristad's (1993) complexity results for language turn out to be, as he conjectures, evidence against informational encapsulation.

[5] Of course, as is most probably the case with paramecia, it is also possible that no behavior is subserved by any Fodorian module at all; see our discussion of qualitative vs quantitative stigmergy in the main body of the text.

[6] In fact, as we pointed out in note 4, according to Fodor (2000) it is almost certainly the case that minds, human ones included, are not massively modular.

[7] In the following discussion, and for the sake of simplicity, we will group PDP/connectionist models and dynamical models together under the single label of "dynamic systems." We are conscious that they are not exactly the same thing although they share a number of basic assumptions, but for the purposes of the

Our strategy in what follows will be the same as the one we've been adopting throughout this book: We will turn to biology. Thus, instead of starting to develop a number of philosophical and theoretical arguments for or against one or the other option, we shall rather try to find some empirical evidence for what we know about the evolution and development of nervous systems in order to settle the issue. Indeed our aim here is to argue that there is no contradiction in our position and that both approaches are in fact compatible.

What do we know then about the evolution and development of nervous systems? We tend to think about nervous systems as the paradigmatic systems for information conduction and processing, but, as Schmidt-Rhaesa (2007: 95) aptly points out:

> The conduction of information does not depend exclusively on the nervous system, as electrical impulses can also be distributed throughout the body, but then in a more or less untargeted way. Additionally, chemical substances are also used as signaling tools with other systems, for example, as hormones in the circulatory system.

And he concludes that "[n]ervous systems may be regarded as specializations for fast and targeted information conduction."[8] Schmidt-Rhaesa's main point is, of course, that the basic building blocks of information conduction that we find today in nervous systems were already there well before anything like a nervous system even existed. For example, electrical signaling also occurs in plants and a number of genes originally assumed to be specific for the nervous system have been found to be present already in the common ancestor of yeasts and metazoans (Schmidt-Rhaesa 2007: 95 for an overview and references, and Minelli 2009: 206). Similarly, in sponges, which do not possess anything like nervous tissues or cells, information about the environment is conducted through their body by means of electrical signals but also perhaps by some chemicals (Farris 2008). All this suggests a very ancient basis for the analog or quantitative mode of computation that is traditionally associated with the brain (Rubel 1985) and would in principle support the idea that nervous systems are better understood as dynamic systems rather than classical computing machines. But let's not be hasty and anticipate conclusions that we may want to revise later. Let's turn to Cnidaria. Cnidarians and ctenophores are the first species where nerves are found and something like a connectionist dream as their nervous system has a net-like architecture (which is sometimes called a plexus), although, as Schmidt-Rhaesa (2007: 96) points out, "there is a trend in cnidarians and ctenophores in which parts of the net are pronounced in some way, creating diversity within the net." However,

---

arguments to be developed in the text we believe that we can gloss over the differences; see however van Gelder (1998) and the open peer commentary that follows the target paper for details.

[8] Although it is often risky to stick to such finalistic explanations in biology. Thus, as Minelli (2009: 206) points out, the nervous system may simply have played an important scaffolding role in maintaining regional differentiation thoughout the animal's development; see also Deutsch and Le Guyader (1998).

although both electrical and chemical synapses occur in these organisms, they are symmetrical so that impulses can be conducted in both directions, the nerve network of these animals being essentially a sort of interface between their different sensory organs, mostly of the static kind (statocysts) to provide information on the orientation of the body and more rarely, also of the photoreceptive kind (Schmidt-Rhaesa 2007: Ch. 7). Thus, nothing suggests here that anything beyond the nomic pattern of reaction exists or that anything different from a continuous, analog, and dynamic conduction of information is going on.[9]

Nothing resembling a brain is present anywhere until we turn our attention to bilaterian animals. Before we go on, however, it is important to settle a couple of questions concerning brains. First, in comparative zoology, the brain is hardly ever defined as a functional unit but rather as an anatomical one: "I call any accumulation of neurons in the anterior end a 'brain'" (Schmidt-Rhaesa 2007: 99), while other possibly present accumulations of neurons in other areas of the body are called "ganglions;" similarly, the terms Peripheral Nervous System (PNS) and Central Nervous System (CNS) only refer to the relative position of these systems with the body wall, with the PNS being closer to it and the CNS sitting deeper in the body. Second, not all bilaterians have a brain (for example echinoderms and hemichordates do not have one) and within those with a CNS, not all have a cephalized CNS, and, finally, within those with a cephalized nervous system only a few of them have higher brain centers (Schmidt-Rhaesa 2007: 113ff.; Farris 2008: 109), meaning that either several episodes of brain reduction have occurred in the evolution from an ancestral "brainy" urbilaterian (Striedter 2005: 109; Schmidt-Rhaesa 2007: 110ff.; Farris 2008: 108) or, rather brains have emerged independently in several bilaterian lineages as some molecular and gene-expression data appear to suggest (see Minelli 2009: §8.2.4). Be this as it may, however, we can identify a number of trends in the evolution of nervous systems (adapted from Schmidt-Rhaesa 2007: 117):

- From diffuse nets to defined nerves;
- From a system without coordinating centers (brain) to systems with a brain;
- From a PNS to a CNS (although a PNS usually persists in addition to a CNS);
- From an undirected (with bidirectional synapses) to a directed system (with unidirectional synapses);
- From a regular grid (orthogon) to more specialized innervation patterns (in protostomes).

But we cannot conceptualize this as a linear process leading to the cephalized CNS with higher brain centers typical of vertebrates, since these processes most probably took place more than once in the evolutionary history of bilaterian animals. There is

---

[9] And, to answer the question in the heading of this section, no, Cnidaria do not seem to have mental representations or, at least, not representations in the traditional sense of the term.

however an interesting correlation between the emergence of cephalized CNSs with higher brain centers and the presence of anterior organs for the detection of visual and olfactory stimuli. Thus, Minelli (2009: 211) notes:

Within a classification of the animal kingdom accepting a partitioning into 33 phyla, about one-third of these major groups include animals without specialized light-detecting organs, one-third have simple light-sensitive structures, and only one-third have organs of such complexity as to be called eyes. Of these, only six phyla (Cnidaria, Mollusca, Annelida, Onychophora, Arthropoda, and Chordata) include species with image-forming eyes.

The correlation for vision is quite robust as, with the exception of cnidarians,[10] the remaining five phyla are the only ones (together with platyhelminthes) with a cephalized CNS and higher brain centers.[11] As for smell little is known, but areas for processing olfactory (and perhaps gustatory) information have been identified in annelids, onychophorans, and arthropods (Harzsch 2006; Farris 2008). Focusing our attention on arthropods and visual systems, most phylogenies within this lineage rely quite strongly on the organization and structure of the visual system and also suggest a strong correlation between a richer eye architecture and an increase in size of the dedicated areas for visual processing (see Harzsch 2006 for a detailed review). In particular, as reported by Farris (2008) this is particularly clear in three insect orders (Coleoptera, Hymenoptera, and Dictyoptera), where the brain structures traditionally known as mushroom bodies and known to play an important role in the processing of sensory information (mostly visual and olfactory) are larger than in other orders and show a characteristic bipartite "gyrencephalic" structure. According to Farris (2008) these large mushroom bodies are likely to have been acquired independently in each lineage as they are not observed in all species. Thus, within Coleoptera only some species of beetles have them, in Hymenoptera they are characteristic of bees and wasps and some ants, and finally, in Dictyoptera, which include cockroaches, termites, and mantids, they appear to be present only in cockroaches. Farris's (2008) conclusion, after a detailed analysis of the ecology of these insects is that the particular organization of these brain centers correlates strongly with a generalist feeding ecology and/or the extensive use of olfactory and visual cues for navigation.[12]

Now things start to get interesting. Could it be the case that the visual systems of the insect orders with large mushroom bodies comprise a specialized module in the

---

[10] So, do Cnidaria have mental representations after all? We still doubt it. One needs something more than a bunch of eyes (often more than 100 in some medusae!) to qualify as a genuine NNSC.

[11] Most photoreceptors in platyhelminthes are either pigment-cup ocelli or simple flat pigment spots; only some species have lenses (Schmidt-Rhaesa 2007: 135–136).

[12] Interestingly, it does not correlate with eusociality. As Farris (2008: 118) states: "The apparent universality of large mushroom bodies across four families of Hymenoptera in which social organization is highly variable provides strong evidence against sociality as a driving force in the initial acquisition of large mushroom bodies in these insects."

strong sense in which we defined modules above? We would like to suggest that this might be the case.

A first datum that could support this conclusion comes also from Farris (2008: 110), who notes that "for the majority of protostomes with centralized nervous systems, the site of multimodal sensory integration, higher computations involved in learning and other forms of plasticity, and in motor planning and execution, is the so-called 'diffuse' neuropil of the protocerebrum," which suggests that mushroom bodies "provide novel processing functions in parallel to or in sequence with protocerebral circuits." To be sure, this is not a definitive argument and we should try to find some additional evidence for our contention.

A possible source for this could be some recent studies in computer modeling of nest building by social insects carried out by Guy Theraulaz and colleagues (Theraulaz and Bonabeau 1995; Theraulaz et al. 1998; Camazine et al. 2001), and of net spinning in spiders by Thiemo Krink and Fritz Vollrath (Krink and Vollrath 1998, 1999). We will limit ourselves to nest building in social insects, but we believe that a similar case to the one we will develop here can be made for web spinning; see the referred papers for details.

The aim of Theraulaz and colleagues' work is to disconfirm the traditional hypothesis that individual insects possess a representation of the global structure to be built and that they make decisions on the basis of that representation. Rather, they set out to demonstrate that "[c]onstruction can best be seen as a morphogenetic process during which past construction sets the stage for new building actions" (Theraulaz et al. 1998: 16). The authors rely on the notion of stigmergy, originally defined by Grassé (1959) to describe the collective process of nest building in termites, and which essentially consists of the idea that a single, simple action performed by an individual results in a small modification of the environment (in this case the structure being built) that influences the actions of the other individuals participating in the process.[13] The net result is a collectively coordinated process with no actual plan or blueprint, where the action of each individual participant in the building is executed on the basis of purely local information. However, as pointed out by Theraulaz and Bonabeau (1995: 386), "[e]ach agent or insect is able to build alone a complete architecture," meaning that the regularity of the emerging structure is independent of the number of agents participating in the building process, and is just a function of the simple stigmergic algorithm followed by each of the individual builders. Now, the notion of stigmergy used by Theraulaz and Bonabeau (1995) is quite general but sufficient to show that a swarm of simple agents is capable of building quite impressive structures, not too different from those observed in nature, by just following a simple algorithm (in fact, a deterministic finite-state automaton)

---

[13] Stigmergy is thus a rough synonym of Wilson's (1975) notion of sematectonic communication.

such that for every configuration c_i of a finite set C of configurations, c_i will trigger a building action a_i from the finite set A of building actions. Of course, the fact the behavior can be modeled by a finite-state automaton does not mean that it is actually controlled by such a computational system. The authors however, don't hesitate at the time of writing in ascribing some reality to their models; let's quote them in full (Theraulaz and Bonabeau 1995: 394):

[O]ur model corresponds to a particular, coarse-grained level of description in which no hypothesis need be made concerning the actual microscopic mechanisms governing the behaviors of the agents: it seems that real insects just act as if they were following some rules, and other levels of description may be more relevant to understand the detail of this behavior. It may be for instance that real wasps do not follow discrete, qualitative rules but are sensitive to, say, quantitative chemical cues. Such levels of description are not in contradiction but are simply compatible with ours, although they are different and can certainly explain different things.

Note that, in essence, the authors claim that a qualitative macrostructure is not necessarily incompatible with a quantitative microstructure, or in other words that the latter implements the former. We wouldn't be surprised if this turned out to be the correct characterization of Fodorian modules—emergent properties of some dynamic system operating at the microstructural level and following different principles from the ones traditionally attributed to classical computational architectures. The crux of the matter is whether there is any empirical evidence of this. Again, we believe, the answer may be found in the work of Theraulaz and his colleagues. In a later paper (Theraulaz et al. 1998) and on the basis of empirical evidence coming from the comparative analysis of nest building in termites and wasps, the authors find themselves in need of refining the notion of stigmergy and distinguishing between quantitative stigmergy and qualitative stigmergy, respectively corresponding to two different mechanisms, self-organization and self-assembly, for coordinating and regulating collective building. The reader is referred to Theraulaz et al.'s paper for the details (see also Camazine et al. 2001: Chs. 17 and 18, for ants and termites, and Ch. 19 for wasps), but the key difference between one and the other mechanism is summarized in the following quote (Theraulaz et al. 1998: pp. 21–22; emphasis in the original):

When termites build pillars, they respond to *quantitative* stimuli, namely, pheromone fields and gradients. Self-assembly is based on a discrete set of stimulus types: For example, an insect responds to a type-1 stimulus with action A and responds to a type-2 stimulus with action B. In other words, qualitatively different stimuli result in qualitatively different responses.

We now invite the reader to compare this characterization of quantitative vs qualitative stigmergy with the nomic/nonnomic distinction that Fodor invoked to distinguish between creatures without mental representation and creatures with mental representation. The parallelism is, in our opinion, quite striking and invites an interpretation of these data in the sense that termites are NSC whereas wasps are NNSC, *at least as far as their nest-building module is concerned*. Thus, the behavior of termites and wasps can in both cases be modeled by a deterministic finite-state automaton, but in the case of wasps it appears that the model actually corresponds to a real computational level of explanation at which operations are computations over mental representations. The point is that the properties to which wasps react (e.g., "being surrounded by three walls") are in all respects the nonnomic kind and cannot just be derived by simple transduction of some specific physical stimulus. In other words some kind of process of perceptual inference appears to be going on in which the said properties are represented in some kind of language. Note that this reasoning is not really different from the kinds of arguments that Fodor has been offering as a defense of the computational nature of modules and for the need to assume some representational language. Moreover this conclusion is perfectly compatible with the idea that other behaviors in wasps, e.g., navigation, are not subserved by classical computational modules. Thus for example, Möller (2000) argues that visual homing strategies in bees and wasps appear to be better captured by an analog processing mechanism than a digital one, but—let us insist on this point once more—there is no contradiction here even if sight is involved both in building and in navigation; on the contrary, this is precisely what is expected of Fodorian modules, that their specialization crosscuts the capacities of the traditional senses (see Fodor 1983: §III.1).

To summarize our position then, we believe that there is no real contradiction between assuming a dynamic perspective to cognition and characterizing certain faculties or modules in a classical way at the same time, as the latter could be emergent properties of the underlying microstructure that arise when the system reaches some critical point. Indeed, this appears to be another trend in the evolution of nervous systems: The emergence of digital, discrete, serial computational systems capable of subserving certain, highly specialized tasks such as nest building in wasps, weaving in some birds, and speaking in humans, giving rise to nonnomic sensitive behaviors which can nevertheless coexist with nomic sensitive ones following an analog, continuous, and parallel regime. Crucially this was clearly not a once-and-for-all invention, as animal phylogenies show us that this may have occurred several times in several lineages and may even have been lost in many others; it did not happen for any particular reason or other either, it just came to be, and tends to be preserved perhaps because, as appears to be the case, a mixed system, combining both analog and digital computations is much more efficient than using just one or the other mode (Sarpeshkar 1998, 2009),[14] a speculation, by the way, that may have

---

[14] For example, Sarpeshkar (2009: 251) writes: "Rather than operate in a collective fashion, with several low-precision digital elements that collectively interact to implement a high-precision or complex

been confirmed by the recent discovery that neurons, at least in mammals, do seem to combine both modes of processing (Alle and Geiger 2006; Shu et al. 2006).[15]

Anyway, these, as we advanced at the beginning, are just speculations. They might be confirmed someday, or proved to be utterly false. This we don't know. But there is something we are sure of: The kinds of questions that will eventually prove or disconfirm them will for certain not be of the "What is it for?" kind. They will be questions asking things like "How does it work?" or "How did it develop?" Daniel Dennett (1987: 316) once wrote that "Some biologists [...] are tempted to renounce *all* talk of function and purpose, and they are right about one thing: there is no stable intermediate position." We couldn't agree more and we hope to have shown that the biologists Dennett mentions are right about another thing: It is good, even necessary, to renounce all talk of function and purpose. Ironically functionalism is not good for much.

---

operation, biological systems operate with many low-precision analog elements that collectively interact to implement a high-precision or complex operation. The interactions and processing can have a hybrid analog-digital nature with both all-or-none digital and graded analog processes being present."

[15] In a very recent paper, Zylberberg et al. (2011) have presented a general neuropsychological model for the nervous system that combines both serial and parallel modes of computation in a way that is quite congenial to our speculations here.

Appendix: On Complexity Issues

One of the central theses of this book is that homologies may be established at the level of computational phenotypes, understood as abstract characterizations of the activity of the natural computational systems implemented in the central nervous systems of most animals. Computational homologies are therefore homology relations among the models of computation that may be realized in the said nervous systems "under every variety of form and function" (Owen 1843). Since in our characterization of computational phenotypes we have been appealing to data about computational complexity coming from two different sources—the Chomsky Hierarchy and the theory of computational complexity, we would like to devote this appendix to present a more detailed exposition of these theories, focusing, in particular, on what specific aspects of computation they concentrate on, how the two relate to each other, and how they are applied in our model.

A very important point to be made from the outset is that both theories are concerned with the study of "languages," which is a technical term in mathematics referring to sets (also in the technical sense) represented in a particular format. A "language" in this sense is a (possibly infinite) set of strings, where a "string" is a concatenation of symbols taken from a finite alphabet. The motivation for this focus on languages is that any set we can think of may be represented in language form, thereby facilitating the study of its properties for reasons that should become clear as the following presentation unfolds. Thus, for example the set of integers $\mathbb{Z} = \{\ldots, -1, 0, 1, \ldots\}$ may be represented in binary by the language made up of strings over the alphabet $\Sigma = \{0, 1\}$ as $\mathbb{Z} = \{\ldots, 10000001, 00000000, 00000001, \ldots\}$, for an 8-bit system.

Also, both theories are interested in classifying languages into equivalence classes and to find out how these classes relate to each other. Complexity classes are therefore sets of languages sharing some specific property. Classes are organized hierarchically such that both the Chomsky Hierarchy and the hierarchy of computational complexity are in fact sets of classes, and one important subject of study in both theories is to find out answers to such questions as: Is class A a subset of class B? Given that we know that class C is a subset of class D, is this subset relation proper or is it the case that both classes are equal? By offering definite answers to these questions, it is possible to have a better understanding of how each hierarchy is structured. Of course, each theory focuses on different properties of languages in order to determine their membership in a specific class and, thence, their specific position in the hierarchy. We shall discuss these in turn, starting with the Chomsky Hierarchy.

The Chomsky Hierarchy is a hierarchy established on the basis of the structural complexity of languages. Thus, the traditional four classes, hierarchically organized as Type3 ⊂ Type2 ⊂ Type1 ⊂ Type0, define four degrees of increasing structural complexity, with the Type3 class being the lowest and Type0 being the highest. Recall that complexity classes are just sets of languages and that languages are themselves sets of strings defined over finite alphabets—there is no direct indication of the complexity of some language in its representation as a set of strings. Chomsky's insight was that one can nonetheless determine the complexity of languages by studying the properties of those finite devices capable of generating them or, in other words, by setting the minimal requirements for a model of computation to be capable of generating some specific type of language. Chomsky defined two equivalent models of computation appropriate for this: Grammars and automata. Without getting into their formal definitions, one can think of grammars as finite sets of rules that, applied in a sequential manner, eventually yield a string of the language. Automata, on the other hand, are abstract machines that can be run such that after a number of well-defined steps they will output a string of the language. Alternatively, an automaton may be seen not as a generator, but as an acceptor, such that if it is given an arbitrary string it will output a "yes" or a "no," signaling whether the string belongs to the language in question or not.[1]

Chomsky also showed that in order to determine the basic structural differences that set apart the languages belonging to different classes in the hierarchy it was just necessary to introduce minimal adjustments to their respective models of computation. Thus, in the case of grammars, Chomsky demonstrated that by gradually relaxing the constraints on the format of rules, it became possible to generate languages of increasing complexity. Now we need to introduce some formal details to make this more explicit. The general format of a grammar rule is $x \to y$, where x and y are strings of symbols, such that the interpretation of the rule is something like the following: "At this step of the derivation of the string, replace substring x with substring y." Now, grammars have two types of symbols, terminals and nonterminals, where the set of terminals is the alphabet from which the strings of the language are defined, whereas nonterminals never show up in the final string, since they are just auxiliary symbols that help to drive the derivation. Thus, for example, the rule $S \to bQ$ contains a single nonterminal on its left-hand side and a terminal followed by a nonterminal on its right-hand side, where we follow the established convention of representing nonterminal symbols with capital letters. Now, suppose we are trying to

---

[1] There are other ways to conceptualize acceptance by an automaton without assuming that it actually outputs anything. For example, we can define acceptance as the situation occurring when the automaton enters a final state and halts, while rejection would correspond to either not halting or halting in a non-final state, for example. Later on we will see that we need to refine the notion of rejection and that the alternatives presented here are in fact not equivalent.

derive a string in some language and that we have reached a point in the derivation in which we have the string *aaaS*. Given the rule above, we can apply it to get the string *aaabQ*. The process will only end when all nonterminals have been successfully replaced with terminals. In principle, both the left-hand side and the right-hand side of a rule can be a string made up of any combination of terminals and nonterminals, but, by defining constraints on the composition of the left- and the right-hand strings, we get different types of grammars capable of generating different types of languages; this is summarized in table A.1.

There are a number of important things to note here. Recall, first, that from the mere inspection of a string we cannot decide whether it belongs to a specific set, since the string bears no record of its structural properties. The only way to decide this question in the affirmative or in the positive is by reference to some grammar or automaton. Thus, the question "Does string *s* belong to set C?" must in fact be posed as "Given grammar G or automaton A, does string *s* belong to set C?" This is a decision problem and, as we have just seen, decision problems can only be framed

TABLE A.1. *Constraints on rules of grammars*

Type of language	Constraints on rules
Recursively enumerable	$x \to y$, where x and y are strings of both terminals and nonterminals.
Context-sensitive (recursive)	$x \to y$, $|x| \leq |y|$, where x and y are strings of both terminals and nonterminals.[a]
Context-free	$A \to y$, where A is a single nonterminal and y is a string of both terminals and nonterminals.[b]
Regular	$A \to wB$, $A \to w$, where A and B are nonterminals and w is a string of terminals.

Type 0 grammars have no constraints on their rules, hence the term "unrestricted" often used to refer to them; context-sensitive grammars require that the left-hand side string be no longer than the right-hand side one; context-free grammars require that the left-hand side be a single nonterminal; regular (also left- or right-linear) grammars additionally require that the right-hand side be either a string of terminals or a string of terminals followed by a single nonterminal (or in the reverse order if left-linear).

[a] Strictly speaking, this constraint defines the rules for grammars capable of generating any recursive set. Context-sensitive grammars would add the following, slightly stronger, constraint: $xAy \to xwy$, i.e., that the substring made up of the single nonterminal A rewrites as the substring w; note that the (possibly null) substrings x and y remain constant on both sides of the arrow and thus define the "context" in which A is to be rewritten as w, hence the term context-sensitive grammar.

[b] In other words, a context-free rule is the result of requiring that the substrings defining the context in a context-sensitive rule be null. Any context-free grammar can moreover be normalized to the much more familiar format $A \to BC$, $A \to a$, where A, B, and C are nonterminals and a is a single terminal, such that there are rules introducing only nonterminal symbols and rules introducing only terminal symbols. Grammars in this format are said to be in Chomsky Normal Form.

with reference to some specific model of computation; as we'll soon also see, decision problems form the core of the theory of computational complexity, but we'd better not anticipate too many things yet. Let's go back to grammars and automata.

Thus, if the structural complexity of a set can only be assessed with respect to one or another model of computation, it should not come as a surprise that Chomsky has always insisted on the fact that what is important in the study of a language (*any* language, natural languages included) is its grammar, not just the strings that make it up. This observation is at the core of the whole generative linguistics enterprise, since, as also pointed out by Chomsky, there are many possible different grammars one can think of capable of generating exactly the same stringset; we shall say in this case that these grammars are all "weakly equivalent" or that they have the same "weak generative capacity." The challenge, when our focus of interest is natural language, is, therefore, which of all the imaginable weakly equivalent grammars is the one that really captures the actual structure of natural language expressions or, using Chomsky's own words, which grammar is the descriptively adequate one. Note that descriptive adequacy is defined not just with respect to weak generative capacity but, rather, with respect to a stronger condition incorporating the notion of structural description. Thus, two weakly equivalent grammars are not necessarily also "strongly equivalent," i.e., they do not necessarily assign the same structural descriptions to the strings of a set or have the same "strong generative capacity." Note that the notion of strong generation somehow transcends the notion of model of computation, since grammars and automata, as defined in formal language theory, are only weak generators, not strong ones; they do not assign structural descriptions to strings in a set.[2] Importantly, though, this does not make all the complexity results presented in the preceding discussion irrelevant, because another crucial outcome of Chomsky's original work is that, whatever the descriptively adequate grammar for natural language eventually turns out to be, its power will be beyond that of context-free grammars, and, as shown in later work already reported in other chapters of this book, it most probably lies within the power of a mildly context-sensitive one. Therefore, the Chomsky Hierarchy sets a minimal lower bound of complexity for any language—also for natural language, which is our model of reference for other complexity results used in this book.

---

[2] Thus, our discussion in Chapter 5, Section 5.2, of the differences between the strings $a_i a_j a_k b_k b_j b_i$ and $a_i a_j a_k b_i b_j b_k$ dealt with issues of strong generative capacity. Indeed, from the point of view of weak generative capacity both strings are the same, but the alternative structural assignment suggested by the subindices in the second string could only be captured by a grammar more powerful than a context-free grammar. Similarly, our discussion at the end of Chapter 7 (Section 7.3) of aural pattern-recognition experiments was motivated by the fact, presented in detail here, that many different grammars, even with different computational power, can weakly generate the same set of strings, so additional information is needed in order to assess exactly what computational model underlies the observed behavior.

That said, it should be clear by now why our computational phenotypes must be interpreted as abstract characterizations of models of computation implemented by natural computational systems. If we wanted to build a performance model of, say, a human's or a Campbell's monkey's verbal behavior, or of a weaverbird's knotting abilities, the model would, among other things, incorporate a descriptively adequate grammar for the system to be able to assign the correct structural descriptions to the corresponding representations or, in other words, to strongly generate the appropriate set of structural descriptions. Strong generation, however, presupposes weak generation and, therefore, a computational phenotype defines the minimal computational architecture for a performance model to be able to account for the complexity of the behavior it subserves.

Thus, along with the Chomsky Hierarchy of *languages*, we can assume that there exist two additional hierarchies, a hierarchy of *grammars* and a hierarchy of *automata*, corresponding to the models of computation capable of generating the languages in the said hierarchy of languages. Another important claim in this book is that models of computation—and, hence, also computational phenotypes—can be differentiated only by reference to the kind of memory regime they incorporate. This is perhaps not entirely obvious in our previous discussion on the different types of grammars, but a closer inspection of the constraints on rules presented in table A.1 reveals that this is in fact the case. Consider, for example, right-linear grammars, a type of regular grammars in which nonterminal symbols can only be introduced as the rightmost symbol of a string made up of just terminals. This constraint has the consequence that each step in the derivation is strictly dependent on the immediately preceding step, such that rules do not carry any information whatsoever of what was done in the preceding steps nor can they condition what will happen two or more steps ahead. If we now turn to context-free grammars, we see that the role of nonterminals is precisely that of carrying information of what has been done in order to determine what will have to be done a number of steps ahead.[3] Similarly with the other grammars, although by letting the rules be more complex, also more complex operations can be carried out. The role of memory is probably also not obvious from the traditional characterization of automata. Thus, even though this is clear in the case of finite-state automata vs pushdown automata, the definition of a linear-bounded automaton for context-sensitive languages is not based on that of the pushdown automaton and this blurs the connection between the different types of automata.

Informally, a finite-state automaton (acceptor version) is defined as a mechanical device with an input tape and a finite control unit. In the former the input string is stored and the latter incorporates a read-only head that scans the input sequentially

---

[3] This is most clearly seen in the demonstration that each context-free grammar is accepted by some pushdown automaton; see the Further reading section at the end of this Appendix for references.

from left to right. We assume that the grammar is somehow hardwired in the control unit and that it drives the unit towards the end of the input string as the head reads the symbols that make it up. If the unit reaches the end of the string and halts in a final state, we say that the automaton has accepted the string. A pushdown automaton is built over this basic architecture with the addition of the pushdown stack and an extra write-erase head capable of manipulating its contents. Thus, as the read-only head scans the input, the write-erase head can push and pop symbols onto and off the stack, respectively, but following a first-in-last-out regime. The conditions for acceptance are similar to those for the finite-state automaton.[4] Now, a linear-bounded automaton is defined as a device with a finite control unit and a single read-write head pointing to the input tape. The unit can both move to the left and to the right and the head can perform any of the two operations on symbols of the input string, either just reading them or overwriting them with another (possibly the same) symbol, with the only constraint being that the movements of the control unit are limited by the length of the input; i.e., the head cannot go to the left of the first symbol nor to the right of the last one. This constraint has the net effect of imposing limitations on the working space of the device and it is in fact a limitation on memory, but it is hard to see the connection with the memory stack of the pushdown automaton, since a linear-bounded automaton is after all a Turing machine with memory limitations.

Our claims concerning memory find however stronger support in the work by David Weir on embedded pushdown automata and their generalization, which defines a progression of increasingly more complex automata where at each stage the degree of nesting of the store is increased to give the next member of the progression. Along with this hierarchy of automata, a hierarchy of languages is defined that is properly contained in the context-sensitive languages and which includes the context-free languages as its structurally simplest family. It is within this context that our claims concerning working memory need to be understood, since here it can more clearly be seen that the structural complexity of a language is a direct function of the sophistication of the storage type associated to the corresponding acceptor, which is always of the pushdown type. Our space of computational phenotypes could then be increased accordingly by including those phenotypes lying between CompuType3 and CompuType0 and equivalent to the corresponding automata within the pushdown hierarchy. It is eventually an empirical question which of these phenotypes would characterize some possible natural computational system and which would, along with CompuType0, fall within the class of theoretically possible but naturally impossible phenotypes. It may be the case, for example,

---

[4] But not exactly the same, since pushdown automata are nondeterministic machines and, unlike finite-state automata, their deterministic version is not equivalent to the nondeterministic one. Deterministic pushdown automata only accept a subfamily of context-free languages, lying properly between regular sets and the rest of the context-free languages.

that constraints on the nature and structure of neural phenotypes impose limitations on the types of storage systems that a brain can implement, but we don't know for the time being whether this is the case or not; this is an open question and a new avenue of research suggested by our proposals in this book.

What we know for sure is that CompuTypeo is impossible because it is equivalent to a Turing machine and Turing machines have no memory limitations; indeed, they have an infinite amount of time and space to do their work. Note, then, that the Chomsky Hierarchy is only concerned with those languages that can be mechanically generated by a Turing machine (or some grammar), but recall that an important point in the study of languages concerns finding out whether a given string is a member in some specific set. Remember also that this can only be established by reference to some model of computation. Now, it happens that this kind of question is not always answerable, and that there is a class of languages within the hierarchy for which the decision problem has no solution. This point deserves some discussion, since here is where the Chomsky Hierarchy and the theory of computational complexity meet each other.

Let us take the Turing machine as our basic model of computation. Now, if we were to start playing the "decision game" by asking some appropriate Turing machine whether some arbitrary string s is a member of some language L in the Chomsky Hierarchy, we would find out that, for some languages, there will be at least a Turing machine M capable of answering "yes" if $s \in L$ and "no" if $s \notin L$. We will say in this case that the machine "decides" the language or that the language is "decidable;" if a language is decidable, we call it a "recursive language." Note that the property of being recursive has an interesting consequence: We can build a Turing machine M' that works in exactly the opposite way to M above, such that it will answer "yes" when $s \notin L$ and "no" when $s \in L$. In other words, machine M' is able to decide the complement set of L, L', which implies that the complement of a recursive set is also recursive, or that recursive languages are closed under complementation—a language is recursive if and only if its complement is also recursive.

We keep on playing the decision game and eventually we come across the following situation: There are certain languages for which a Turing machine is able to answer "yes" if $s \in L$, but that is unable to halt and answer "no" if $s \notin L$. In this case we simply say that the Turing machine "accepts" the language and call the language "recursively enumerable." Note that, by the definition of acceptance just given, any recursive language is also recursively enumerable, but not the other way around—some recursively enumerable languages are not recursive and are in fact undecidable. Now, if a language is not recursive, then, given the set-theoretic definition of recursive, its complement is not recursive either, indeed, it is not even recursively enumerable, because recursively enumerable languages are not closed under complementation: For a language and its complement to be both recursively enumerable, at least one of them has to be recursive.

Thus, there are recursively enumerable languages that are not recursive and, even worse, there are languages that are not even recursively enumerable. This shows that the universe of languages very much transcends that of the Chomsky Hierarchy—there are languages that the Hierarchy does not even contemplate, but it most crucially shows that we can look at languages as if they were "problems." Patently, one thing we want to know about a problem is if it is solvable at all, and we have just seen that some problems are not: The undecidable ones. There are many other things we may want to know about a problem apart from its solvability—we may want to know, for example, how hard it is to solve it. And here enters the theory of computational complexity.

At the beginning of this appendix we saw that this theory is also concerned with sets, and we have just seen that it looks at sets as if they were problems in order to determine how hard it is to solve them. To be sure, one important assumption of the theory of computational complexity is that any problem we can think of may be reduced to a language recognition problem in order to assess its complexity. Thus, the very same languages that fall under the focus of the Chomsky Hierarchy fall also under the focus of the theory of computational complexity, with the additional fact that the latter is also interested in languages within the realm of the undecidable and beyond, since some problems may indeed turn out to be not just hard to solve but actually unsolvable.

There are many ways to assess the hardness of a problem, its complexity, but the most common one in computational complexity is to determine the maximal amount of time and space resources consumed by a Turing machine for its solution. Thus, the theory of computational complexity assumes a fixed *model* of computation with reference to which complexity measures are defined, but not necessarily a fixed *mode* of computation: The same model (the Turing machine) operating in different modes (e.g., deterministic vs nondeterministic) may yield different complexity results, because nondeterministic devices are more powerful than deterministic ones and certain problems that do not find an efficient solution in deterministic mode may have one in nondeterministic mode.

Switching from one to another mode of computation is nevertheless too crude a measure of complexity, especially when the focus is on efficiency. In order to measure the complexity of algorithms, the theory of computational complexity also pays attention to the resources consumed by the device to solve the problem. The most common resources of interest (there are others) are time and space, which, for a Turing machine, are defined respectively as the number of steps to halting and the number of cells in the tape visited during the whole computation. Note, however, that Turing machines are devices with an unlimited amount of resources, meaning that, if a problem is decidable then a Turing machine will eventually halt, but possibly after having spent a much too large amount of time and space. Inasmuch as one is concerned with the efficiency of algorithms one needs hence to find out whether a solution is reachable within some feasible time and/or space bounds. The usual practice in this case (and also the easiest result one can prove for an algorithm) is to determine an upper bound of complexity, i.e., the worst case beyond which efficiency severely decays. It is nonetheless also possible (but also much more difficult) to determine a lower bound of complexity, meaning that if we are able to locate the lower complexity bound for a problem in some class C, then any possible algorithm we might devise for it will always be at least as complex, i.e., that the problem in question is as hard as any of the hardest problems in C (it could be harder, but not easier); in those cases, we say that the problem is C-hard.

Back to complexity measures in terms of time and space, these are defined as functions on the length of the input string telling us the rate at which time or space grows as the length of the input grows, expressed, in the so-called "big oh notation," as $f(n) = O(g(n))$, where n is the length of the input. These functions may be seen as limits where n approaches ∞ (there is no theoretical upper bound for the length of the input) and, consequently, telling us the rate at which time or space resources grow as the length of the input grows. While the rate of growth of the latter is always linear, that of the former need not be linear, but it may rather be polynomial (i.e., $O(n^c)$, c a constant), exponential (i.e., $O(c^n)$, c a constant), or even factorial (i.e., $O(n!)$). Thus, if for example we have an algorithm running in time $O(n^2)$ and another running in time $O(2^n)$, little difference will be observed for small values of n, but it is clear that the exponential one will soon become inefficient as exponential functions grow much faster than polynomial ones. Of course, some functions may identify algorithms showing better values than polynomial ones, such as linear (i.e., $O(n)$), logarithmic (i.e., $O(\log n)$), or constant (i.e., $O(c)$) values. Now remember that, unless we are aiming at solving the more difficult task of setting a lower bound of complexity for some problem, these functions are to be interpreted as worst-case solutions fixing an upper bound of complexity. Thus, and assuming we were studying time bounds, the results for the two algorithms above should be interpreted in the sense that, in the worst case, we will reach a solution in a time equal to the square of the input in the polynomial case, but equal to 2 to the power of the length of the input in the exponential case. We will then classify the corresponding languages as members of the classes TIME(n^2) and TIME(2^{n^1}), respectively.[5] Given that the exponent in a polynomial may be any integer, there are infinitely many polynomial and exponential classes identifying more or less complex languages and defining a hierarchy of complexity within the polynomial and the exponential realm. These two hierarchies, defined as the union of all the polynomial classes on the one hand, and the union of all exponential classes on the other, specify two important complexity classes denoted respectively as **P** and **EXP**. Since **P** and **EXP** are themselves hierarchies, it follows that

---

[5] In exponential classes, it is usually assumed that the base is fixed to 2, while the exponent is in fact a polynomial n^k, such that exponential growth is determined by variation on the value of k.

some problems in **P**, for example, are harder than others. The hardest problems in **P** are the **P**-complete problems, where a C-complete problem, C a complexity class, is a problem that is at least as hard as any of the hardest problems in C (i.e., it is C-hard) and it is known to be in C. Complete problems are very important because they are "model" problems for which both a lower and an upper bound of complexity is known and these are defined by the class in which the problem belongs; in other words, a **P**-complete problem is at least as hard as any other problem in **P** but not harder. Accordingly, if we manage to show that some problem P_n we are studying is equivalent to some known complete problem P_c,[6] we will have an exact and precise characterization in complexity terms of the problem in question, since we will know that P_n is at least as hard as P_c and it may eventually be the case that it is also a complete problem (i.e., in the same class as P_c).

Now, **P** and **EXP** are both deterministic classes, meaning that any problem within these classes can be solved efficiently in, respectively, polynomial or exponential time by a Turing machine working in deterministic mode. Nondeterministic time classes are a bit different. This is mostly due to the fact that nondeterminism is still a poorly understood notion and that the definition of recognition is weaker for nondeterministic Turing machines than it is for deterministic ones. Given the fact that a nondeterministic Turing machine, at any point of the computation, has at least two choices to follow, time measures do not refer to all the possible steps in a single computation (these would be too many) and are calculated differently assuming that at least one path yields to acceptance. The most important nondeterministic time class is **NP** (for Nondeterministic Polynomial) and, like **P**, is defined as the union of all nondeterministic classes with polynomial characteristic functions. The discovery that some problem is in **NP** means that the Turing machine will reach a solution following some path and that the correctness of the solution can be verified through a succinct certificate (or polynomial witness) in polynomial time. The succinct certificate is an external device that can be consulted every time a "yes" state has been reached and that provides an efficient procedure to check the result.

Turning briefly to space complexity classes, these are constructed in exactly the same way as time complexity classes, with the proviso that space is taken to be a more costly resource than time because it can be reused. In the case of space, then, polynomial functions identify very hard problems, close to intractability. For this reason, when dealing with space, logarithmic or linear bounds are preferred over polynomial bounds, although the class that will be of interest for us here is, precisely, **PSPACE** (for Polynomial Space), which is a deterministic class.

---

[6] Where equivalence is demonstrated by the possibility of reducing the complete problem to the problem under study. Reductions are conversions of instances of some problem of known complexity to instances of the problem whose complexity we wish to determine.

Finally, the four classes considered here are related by inclusion, composing the following hierarchy of increasing complexity: $\mathbf{P} \subseteq \mathbf{NP} \subseteq \mathbf{PSPACE} \subseteq \mathbf{EXP}$. Note that the inclusions are not known to be proper, meaning that the classes might turn out to be equal. Indeed, the question whether $\mathbf{P} = \mathbf{NP}$ is one of the most important unsolved problems in complexity theory.

Now that we have a detailed definition of the two complexity hierarchies, the structural and the computational, let us see how the two are connected. Take first the case of natural language, which as we have seen is structurally more complex than context-free but less so than context-sensitive, with a structural complexity equivalent to that of a mildly context-sensitive language. Recall that these complexity results refer to sets and that these very same sets will appear as members of some computational complexity class represented as recognition problems. The problem CONTEXT-FREE RECOGNITION is in **P**; the problem CONTEXT-SENSITIVE RECOGNITION is in **PSPACE**;[7] then, the problem MILD-CONTEXT-SENSITIVE RECOGNITION should fall somewhere in between, or within any of the two bounding classes—a conjecture that finds its confirmation in the results reported in Chapter 7 that language computations are **NP**-complete. Note that these inferences are possible because both theories deal with the very same entities (i.e., languages), but classify them according to different criteria. Therefore, two languages known to be in the same structural class (e.g., context-free) will fall within the same complexity class (i.e., **P**), and vice versa; two languages known to be in the same computational complexity class will necessarily fall within the same structural class. Similarly, if natural language is **NP**-complete and knot recognition is **NP**-complete too, then both problems are equivalent on computational complexity grounds, but, qua languages, they will be represented by sets of an equivalent structural complexity and, hence, each specifiable by an automaton/grammar also of equivalent complexity.

It is thus in this sense that for example our claims concerning the equivalence of language and knot recognition must be interpreted and that this implies the presence of at least a computational phenotype of a specific kind.

Further Reading

In the foregoing presentation we tried to provide an informal and accessible introduction both to formal language and automata theory and to computational complexity theory, hopefully sufficient to understand our proposals concerning computational phenotypes and their homologies. There may be readers, however, willing to go deeper into the details and it is for them that we compiled this brief bibliographical note.

---

[7] It is in fact **PSPACE**-complete.

Chomsky's original work on formal languages and automata is scattered across a number of papers published between 1953 and 1963, but Chomsky (1956a), Chomsky (1959b), and especially Chomsky (1963) remain perhaps as the most complete and comprehensive references. Chomsky (1957) is also a good summary of these results, in particular for his argument against the adequacy of context-free grammars. The issue of weak vs strong generative capacity is discussed at length in Chomsky (1965). For additional information on languages and automata, and the equivalence between grammars and automata, the reader may want to consult Hopcroft and Ullman (1979) and Lewis and Papadimitriou (1981), although the latter focuses mostly on regular and context-free languages. As for the theory of computational complexity, the paper by Fortnow and Homer (2003) offers a short historical survey of its main past and present concerns. For a more complete presentation, Papadimitriou (1994) is an excellent source of information, in particular for all concerning the issues of determinism vs nondeterminism, the relations among different complexity classes and their characterization, and for descriptions of some model problems in computational complexity theory. The differences between the structural or qualitative approach of formal language theory and the quantitative one of complexity theory are discussed in Hopcroft and Ullman (1979: Ch. 12) and Lewis and Papadimitriou (1981: Ch. 7); Chapter 12 of Hopcroft and Ullman's book also contains a good introduction to time and space bounds that may be a good complement to that of Papadimitriou (1994). Finally, an excellent—and updated—text with a clear and comprehensive introduction to both formal language and complexity theories and the connections between each other is Webber (2008).

References

Aboitiz, Francisco, and García, Ricardo (1997). 'The evolutionary origins of the language areas in the human brain. A neuroanatomical perspective'. *Brain Research Reviews* 25: 381–396.

—— —— Brinetti, Enzo, and Bosman, Conrado (2006). 'The origins of Broca's Area and its connection from an ancestral working-memory network'. In Y. Grodzinsky and K. Amunts (eds), *Broca's Region*. Oxford: Oxford University Press, 3–16.

Adams, Colin C. (2004). *The Knot Book. Second edition*. Providence, RI: American Mathematical Society.

Agol, Ian, Hass, Joel, and Thurston, William (2002). '3-manifold knot genus is NP-complete'. In *Proceedings of the Thirty-fourth Annual ACM Symposium on Theory of Computing*. New York: ACM Press, 761–766.

—— —— —— (2005). 'The computational complexity of knot genus and spanning area'. *Transactions of the American Mathematical Society* 358(9): 3821–3850.

Aho, Alfred V. (1968). 'Indexed grammars—An extension of context-free grammars'. *Journal of the ACM* 15: 647–671.

Aitchinson, Jane (1996). *The Seeds of Speech. Language Origin and Evolution*. Cambridge: Cambridge University Press.

Alba, David B. (2002). 'Shape and stage in heterochronic models'. In N. Minugh-Purvis and K. J. McNamara (eds), *Human Evolution through Developmental Change*. Baltimore, MD: The Johns Hopkins University Press, 28–50.

Alberch, Pere (1980). 'Ontogenesis and morphological diversification'. *American Zoologist* 20: 653–667.

—— (1982). 'Developmental constraints in evolutionary processes'. In J. T. Bonner (ed.), *Evolution and Development. Dahlen Konferenzen*. New York: Springer-Verlag, 313–332.

—— (1985). 'Problems with the interpretation of developmental sequences'. *Systematic Zoology* 34: 46–58.

—— (1989). 'The logic of monsters: Evidence for internal constraint in development and evolution'. *Geobios* 12 (mémoire spécial): 21–57.

—— (1991). 'Del gen al fenotipo: Sistemas dinámicos y evolución morfológica'. *Revista Española de Paleontología* (número extraordinario 'El estudio de la forma orgánica y sus secuencias en Paleontología Sistemática, Paleontología y Paleontología Evolutiva'): 13–19.

—— and Alberch, Jordi (1981). 'Heterochronic mechanisms of morphological diversification and evolutionary change in the neotropical salamander, *Bolitoglossa occidentalis* (Amphibia: Plethodontidae)'. *Journal of Morphology* 167: 249–264.

—— and Blanco, María José (1996). 'Evolutionary patterns in ontogenetic transformation'. *International Journal of Developmental Biology* 40: 845–858.

—— Gould, Stephen J., Oster, George F., and Wake, David B. (1979). 'Size and shape in ontogeny and phylogeny'. *Paleobiology* 5: 296–317.

Alexander, Richard D. (1987). *The Biology of Moral Systems*. New York: Aldine de Gruyter.

Alle, Henrik, and Geiger, Jörg R. P. (2006). 'Combined analog and action potential coding in hyppocampal mossy fibers'. *Science* 311: 1290–1293.

Ambrose, Stanley H. (1998). 'Late Pleistocene human population bottlenecks, volcanic winter, and differentiation of modern humans'. *Journal of Human Evolution* 34: 623–651.

Amundson, Ronald A. (2005). *The Changing Role of the Embryo in Evolutionary Thought: Roots of Evo-Devo*. Cambridge: Cambridge University Press.

—— (2006). 'EvoDevo as cognitive psychology'. *Biological Theory* 1: 10–11.

—— and Lauder, George V. (1994). 'Function without purpose: The uses of causal role function in evolutionary biology'. *Biology and Philosophy* 9: 443–469.

Arendt, Detlev (2005). 'Genes and homology in nervous system evolution: Comparing gene functions, expression patterns, and cell type molecular fingerprints'. *Theory in Biosciences* 124: 185–197.

Aristotle (1990). *Historia de los animales*. Madrid: Akal, translated by José Vara Maldonado [*History of Animals* (3 vols.). Ed. D. M. Balme and A. L. Peck. Cambridge, MA: Harvard University Press (Loeb Classical Library 437–439), 1965, 1970, 1991].

—— (1999). *Acerca del alma*. Madrid: Gredos, translated by Tomás Calvo Martínez [*On the Soul*. Ed. Hugh Lawson-Tancred. London: Penguin, 1987].

—— (2000). *Partes de los animales. Marcha de los animales. Movimiento de los animales*. Madrid: Gredos, translated by Elvira Jiménez Sánchez-Escariche and Almudena Alonso Miguel [*Parts of Animals. Progression of Animals. Movement of Animals*. Ed. A. L. Peck and E. S. Forster. Cambridge, MA: Harvard University Press (Loeb Classical Library 323), 1937].

Armstrong, David M. (1968). *A Materialist Theory of the Mind*. London: Routledge.

Arnold, Kate, and Zuberbühler, Klaus (2006a). 'Semantic combinations in primate calls'. *Nature* 441: 303.

—— —— (2006b). 'The alarm-calling system of adult male putty-nose monkeys *Cercopithecus nictitans martini*'. *Animal Behaviour* 72: 643–653.

Arsuaga, Juan Luis (1999). *El collar del neandertal*. Madrid: Temas de Hoy [*The Neanderthal's Necklace*. Tr. Andy Klatt. New York: Basic Books, 2002].

Arthur, Wallace (2000). 'The concept of developmental reprogramming and the quest for an inclusive theory of evolutionary mechanisms'. *Evolution & Development* 2(1): 49–57.

—— (2004). *Biased Embryos and Evolution*. Cambridge: Cambridge University Press.

—— (2011). *Evolution. A Developmental Approach*. Oxford: Wiley-Blackwell.

Atkinson, Quentin D. (2011). 'Phonemic diversity supports a serial founder effect model of language expansion from Africa'. *Science* 332: 346–349.

Atz, James W. (1970). 'The application of the idea of homology to behavior'. In L. R. Aronson, E. Tobach, D. S. Lehrman, and J. S. Rosenblatt (eds), *Development and Evolution of Behavior*. San Francisco: Freeman, 53–74.

Avian Brain Nomenclature Consortium (2005). 'Avian brains and a new understanding of vertebrate brain evolution'. *Nature Reviews* 6: 151–159.

Avital, Eytan, and Jablonka, Eva (2000). *Animal Traditions. Behavioural Inheritance in Evolution*. Cambridge: Cambridge University Press.

Baddeley, Alan (2007). *Working Memory, Thought, and Action*. Oxford: Oxford University Press.

Baker, Mark C. (1996). *The Polysynthesis Parameter*. Oxford: Oxford University Press.

Balari, Sergio (2005). 'Desarrollo y complejidad computacional: ¿Dos elementos clave para comprender los orígenes del lenguaje?'. *Ludus Vitalis. Revista de Filosofía de las Ciencias de la Vida* XIII(24): 181–198.

—— (2006). 'Heterochronies in brain development and the origins of language. A coevolutionary scenario'. In J. Roselló and J. Martín (eds), *The Biolinguistic Turn. Issues on Language and Biology*. Barcelona: PPU, 66–81.

—— Benítez-Burraco, Antonio, Camps, Marta, Longa, Víctor M., Lorenzo, Guillermo, and Uriagereka, Juan (2008). '¿Homo loquens neanderthalensis? En torno a las capacidades simbólicas y lingüísticas del Neandertal'. *Munibe (Atropologia-Arkeologia)* 59: 3–24.

—— —— —— —— —— (2011). 'The archaeological record speaks: Bridging Anthropology and Linguistics'. *International Journal of Evolutionary Biology*, volume 2011 (Special Issue: Key Evolutionary Transformations in *Homo sapiens*), Article ID 382679, 17 pages, doi:10.4061/2011/382679.

—— —— Longa, Víctor M., and Lorenzo, Guillermo (in press). 'The fossils of language: What are they, who has them, how did they evolve?'. In C. Boeckx and K. Grohmann (eds), *The Cambridge Handbook of Biolinguistics*. Cambridge: Cambridge University Press.

—— and Lorenzo, Guillermo (2008). 'Pere Alberch's developmental morphospaces and the evolution of cognition'. *Biological Theory* 3(4): 297–304.

—— —— (2009). 'Computational phenotypes: Where the theory of computation meets Evo-Devo'. *Biolinguistics* 3(1): 2–60.

—— —— (2010a). '¿Para qué sirve un ballestrinque? Reflexiones sobre el funcionamiento de artefactos y organismos en un mundo sin funciones'. *Teorema. Revista Internacional de Filosofía* 29(1): 56–76.

—— —— (2010b). 'Specters of Marx. A review of *Adam's Tongue* by Derek Bickerton'. *Biolinguistics* 4(1): 115–126.

—— —— (2010c). 'Incidental biology. A reply to Derek Bickerton's response'. *Biolinguistics* 4(1): 132–136.

—— —— (2012). 'Reivindicación del villano'. In R. Owen, *Discurso sobre la naturaleza de las extremidades* (S. Balari and G. Lorenzo, eds.). Oviedo: KRK.

Barham, Lawrence S. (2002). 'Systematic pigment use in the Middle Pleistocene of South-central Africa'. *Current Anthropology* 31: 181–190.

Bar-Hillel, Yehoshua (1955). 'An examination of information theory'. *Philosophy of Science* 22: 86–105.

Barlow, George W. (1977). 'Modal action patterns'. In T. A. Sebeok (ed.), *How Animals Communicate*. Bloomington, IN: Indiana University Press, 98–134.

Barret, H. Clark (2011). 'The wrong kind of wrong: A review of *What Darwin Got Wrong*'. *Evolution and Human Behavior* 32: 76–78.

Barton, Robert A., and Harvey, Paul H. (2000). 'Mosaic evolution of brain structure in mammals'. *Nature* 405: 1055–1058.

Bateson, Gregory (1949). 'Bali: The value system of a steady state'. In M. Fortes (ed.), *Social Structure: Studies Presented to A. R. Radcliffe-Brown*. Oxford: Clarendon Press, 35–53.

Bateson, Gregory (1972). 'The logical categories of learning and communication'. In *Steps to an Ecology of Mind*. Chicago: The University of Chicago Press: 279–308.

Bateson, William (1894). *Materials for the Study of Variation Treated with Especial Regard to Discontinuity in the Origin of Species*. London: Macmillan.

Bayman, Benjamin F. (1977). 'Theory of hitches'. *American Journal of Physics* 84: 185–190.

Behar, Doron M., Villems, Richard, Soodyall, Himla, Blue-Smith, Jason, Pereira, Luisa, Metspalu, Ene, Scozzari, Rosaria, Makkan, Heeran, Tzur, Shay, Comas, David, Bertranpetit, Jaume, Quintana-Murci, Lluís, Tyler-Smith, Chris, Wells, R. Spencer, Rosset, Saharon, and The Genographic Consortium (2008). 'The dawn of human matrilineal diversity'. *American Journal of Human Genetics* 82(5): 1130–1140.

Benítez-Burraco, Antonio (2009). *Genes y Lenguaje. Aspectos Ontogenéticos, Filogenéticos y Cognitivos*. Barcelona: Reverté.

—— and Longa, Víctor M. (2010). 'Evo-Devo—Of course, but which one? Some comments on Chomsky's analogies between the biolinguistic approach and Evo-Devo'. *Biolinguistics* 4(4): 308–323.

—— —— Lorenzo, Guillermo, and Uriagereka, Juan (2008). 'Also sprach Neanderthalis... or did she?'. *Biolinguistics* 2(2–3): 225–232.

Benveniste, Émile (1970). 'L'appareil formel de l'énonciation'. *Langages* 17: 12–18.

Berwick, Robert C., and Chomsky, Noam (2011). 'The Biolinguistic Program: The current state of its development'. In A. M. Di Sciullo and C. Boeckx (eds), *The Biolinguistic Enterprise. New Perspectives on the Evolution and Nature of the Human Language Faculty*. Oxford: Oxford University Press, 19–41.

—— Okanoya, Kazuo, Beckers, Gabriel J. L., and Bolhuis, Johan J. (2011). 'Songs to syntax: The linguistics of birdsong'. *Trends in Cognitive Sciences* 15: 113–121.

Bickerton, Derek (1990). *Language and Species*. Chicago: The University of Chicago Press.

—— (2009). *Adam's Tongue: How Humans Made Language, How Language Made Humans*. New York: Hill and Wang.

Biederman, Irving (1987). 'Recognition-by-Components: A theory of human image understanding'. *Psychological Review* 94(2): 115–147.

Blake, Barry J. (2001). *Case*. Cambridge: Cambridge University Press.

Block, Ned, and Kitcher, Philip (2010). 'Misunderstanding Darwin. Natural selection's secular critics get it wrong'. *Boston Review*, March/April. <http://bostonreview.net/BR35.2/block_kitcher.php>.

Bock, Gregory R., and Cardew, Gail (eds) (1999). *Homology*. Chichester: John Wiley & Sons/Novartis Foundation.

Boeckx, Cedric (2009). *Language in Cognition. Uncovering Mental Structures and the Rules Behind Them*. Malden, MA: Wiley-Blackwell.

—— (2010). 'What Principles and Parameters got wrong'. Ms., Universitat Autònoma de Barcelona.

—— and Grohmann, Kleanthes (2007). 'The Biolinguistics Manifesto'. *Biolinguistics* 1: 1–8.

Bonato, Lucio, and Minelli, Alessandro (2002). 'Parental care in *Dicellophilus carniolensis* (C. L. Koch, 1847): New behavioral evidence with implications for the higher phylogeny of centipedes (Chilopoda)'. *Zoologischer Anzeiger* 241: 193–198.

Bond, Jacquelyn, Roberts, Emma, Mochida, Ganesh H., Hampshire, Daniel J., Scott, Sheila, Askham, Jonathan M., Springell, Kelly, Mahadevan, Meera, Crow, Yanick J., Markham, Alexander F., Walsh, Christopher A., and Woods, C. Geoffery (2002). 'ASPM is a major determinant of cerebral cortical size'. *Nature Genetics* 32: 316–320.

Bonner, John Tyler (ed.) (1982). *Evolution and Development. Dahlen Konferenzen*. New York: Springer-Verlag.

Borgia, Gerald (1986). 'Sexual selection in bowerbirds'. *Scientific American* 254: 70–79.

Botha, Rudolf (2009). 'Theoretical underpinnings of inferences about language evolution: The syntax used at Blombos Cave'. In R. Botha and C. Knight (eds), *The Cradle of Language*. New York: Oxford University Press, 93–111.

Bourikas, Dimitris, Pekarik, Vladimir, Baeriswyl, Thomas, Grunditz, Asa, Sadhu, Rejina, Nardó, Michelle, and Stoeckli, Esther T. (2005). 'Sonic hedgehog guides commissural axons along the longitudinal axis of the spinal cord'. *Nature Neuroscience* 8: 297–304.

Bouzouggar, Abdeljalil, Barton, Nick, Vanhaeren, Marian, d'Errico, Francesco, Collcut, Simon, Higham, Tom, Hodge, Edward, Parfitt, Simon, Rhodes, Edward, Schwenninger, Jean-Luc, Stringer, Chris, Turner, Elaine, Ward, Steven, Moutmir, Abdelkrim, and Stambouli, Adelhamid (2007). '82,000-year-old shell beads from North Africa and implications for the origins of modern human behavior'. *Proceedings of the National Academy of Sciences USA* 104: 9964–9969.

Boyden, Alan (1947). 'Homology and analogy. A critical review of the meanings and implications of these concepts in biology'. *The American Midland Naturalist* 37: 548–241.

—— (1969). 'Homology and analogy'. *Science* 164: 455–456.

Braitenberg, Valentino (1984). *Vehicles. Experiments in Synthetic Psychology*. Cambridge, MA: MIT Press.

Brigandt, Ingo (2002). 'Homology and the origin of correspondence'. *Biology and Philosophy* 17: 389–407.

—— (2009). 'Natural kinds in evolution and systematics: Metaphysical and epistemological considerations'. *Acta Biotheoretica* 57: 77–97.

Brockmann, H. Jane, Grafen, Alan, and Dawkins, Richard (1979). 'Evolutionary Stable Strategy in a digger wasp'. *Journal of Theoretical Biology* 77: 473–496.

Brown, William M. (2008). 'Sociogenomics for the cognitive adaptationist'. In Ch. Crawford and D. Krebs (eds), *Foundations of Evolutionary Psychology*. New York: Lawrence Erlbaum: 117–135.

Bullock, Theodore H., and Horridge, G. Adrian (1965). *Structure and Function in the Nervous System of Vertebrates*. San Francisco: W.H. Freeman.

Burish, Mark J., Kueh, Hao Yuan, and Wang, Samuel S.-H. (2004). 'Brain architecture and social complexity in modern and ancient birds'. *Brain, Behavior and Evolution* 63(2): 107–124.

Buss, David M. (2007). *Evolutionary Psychology: The New Science of the Mind*. Boston: Allyn and Bacon,

Byers, John A. (1992). 'The mobility gradient: Useful, general, falsifiable?'. *Behavioral and Brain Sciences* 15: 270–271.

Camazine, Scott, Deneubourg, Jean-Louis, Franks, Nigel R., Sneyd, James, Theraulaz, Guy, and Bonabeau, Eric (2001). *Self-Organization in Biological Systems*. Princeton, NJ: Princeton University Press.

Camps, Marta, and Uriagereka, Juan (2006). 'The Gordian Knot of linguistic fossils'. In J. Rosselló and J. Martín (eds), *The Biolinguistic Turn. Issues on Language and Biology*. Barcelona: PPU, 34–65.

Canguilhem, Georges (1965). *La connaissance de la vie*. Paris: Vrin.

—— (1966). *Le normal et le pathologique*. Paris: PUF.

Cardín, Alberto (1994). *Dialéctica y canibalismo*. Barcelona: Anagrama.

Carroll, Sean B. (2005). *Endless Forms Most Beautiful: The New Science of Evo Devo and the Making of the Animal Kingdom*. New York: W.W. Norton & Company.

—— (2008). 'Evo-Devo and an expanding evolutionary synthesis: A genetic theory of morphological evolution'. *Cell* 134: 25–36.

Casati, Roberto, and Varzi, Achille C. (1999). *Parts and Places. The Structures of Spatial Representation*. Cambridge, MA: MIT Press.

Cave, Carolyn Baker, and Kosslyn, Stephen M. (1993). 'The role of parts and spatial relations in object identification'. *Perception* 22: 229–248.

Chalmers, David J. (1996). *The Conscious Mind. In Search of a Fundamental Theory*. Oxford: Oxford University Press.

Cheney, Dorothy L., and Seyfarth, Robert M. (1990). *How Monkeys See the World. Inside the Mind of Another Species*. Chicago: The University of Chicago Press.

Chierchia, Gennaro (1998). 'Plurality of mass nouns and the notion of semantic parameter'. In S. Rothstein (ed.) *Events and Grammar*. Dordrecht: Kluwer Academic Publishers, 53–103.

Chihara, Charles, and Fodor, Jerry A. (1965). 'Operationalism and ordinary language'. *American Philosophical Quarterly* II: 281–295 (Reprinted in Fodor 1981: 35–62).

Ching, Yick-Pang, Qi, Robert Z., and Wang, Jerry H. (2000). 'Cloning of three novel neuronal Cdk5 activator binding proteins'. *Gene* 242: 285–294.

Chomsky, Noam (1956a). 'Three models for the description of language'. *IRE Transactions on Information Theory* 2: 113–124.

—— (1956b). 'On the limits of finite-state description'. *Quarterly Progress Report* 42: 64–65.

—— (1957). *Syntactic Structures*. The Hague: Mouton.

—— (1959a). 'Review of *Verbal Behavior* by B. F. Skinner'. *Language* 35: 26–58.

—— (1959b). 'On certain formal properties of grammars'. *Information and Control* 2: 137–167.

—— (1963). 'Formal properties of grammars'. In R. D. Luce, R. R. Bush, and E. Galanter (eds), *Handbook of Mathematical Psychology, Vol. 2*. New York: Wiley, 323–418.

—— (1965). *Aspects of the Theory of Syntax*. Cambridge, MA: MIT Press.

—— (1966). *Cartesian Linguistics. A Chapter in the History of Rationalist Thought*. New York: Harper and Row.

—— (1968). *Language and Mind*. New York: Harcourt Brace Jovanovich.

—— (1975). *Reflections on Language*. New York: Pantheon Books.

—— (1980). *Rules and Representations*. New York: Columbia University Press.

—— (1986). *Knowledge of Language. Its Nature, Origins, and Use*. New York: Praeger.

—— (1993). *Language and Thought*. Wakefield, RI: Moyer Bell.

—— (1995). *The Minimalist Program*. Cambridge, MA: MIT Press.

—— (2000). 'Minimalist inquiries: The framework'. In R. Martin, D. Michaels, and J. Uriagereka (eds), *Step by Step. Essays in Honor of Howard Lasnik*. Cambridge, MA: MIT Press, 89–115.

—— (2004). 'Beyond explanatory adequacy'. In A. Belletti (ed.), *Structures and Beyond. The Cartography of Syntactic Structures*. Oxford: Oxford University Press, 104–131.

—— (2005). 'Three factors in language design'. *Linguistic Inquiry* 36: 1–22.

—— (2010). 'Some simple evo devo theses: How true might they be for language'. In R. K. Larson, V. Déprez, and H. Yamakido (eds), *The Evolution of Language. Biolinguistic Perspectives*. Cambridge: Cambridge University Press, 45–62.

Christy, John H. (1988). 'Pillar function in the fiddler crab *Uca beebei*. II: Competitive courtship signaling'. *Ethology* 78: 113–128.

Churchland, Paul M. (1981). 'Eliminative materialism and the propositional attitudes'. *Journal of Philosophy* 78: 67–90.

—— (1985). 'Reduction, qualia and the direct inspection of brain states'. *Journal of Philosophy* 82: 8–28.

—— (1995). *The Engine of Reason, the Seat of the Soul*. Cambridge, MA: MIT Press.

Clark, Andy (1997). *Being There. Putting Brain, Body and World Together Again*. Cambridge, MA: MIT Press.

—— and Chalmers, David J. (1998). 'The extended mind'. *Analysis* 58(1): 7–19.

Claudius Aelianus [Aelian] (1989). *Historia de los animales*. Madrid: Akal, translated by José Vara Maldonado [*On Animals* (3 vols.). Ed. A. F. Scholfield. Cambridge, MA: Harvard University Press (Loeb Classical Library 446, 448, 449), 1958–1959].

Collias, Elsie C., and Collias, Nicholas E. (1973). 'Further studies on development of nest-building behavior in a weaverbird (*Ploceus cucullatus*)'. *Animal Behavior* 21: 371–382.

Collias, Nicholas E., and Collias, Elsie C. (1962). 'An experimental study of the mechanisms of nest building in a weaverbird'. *The Auk* 79: 568–595.

Collins, Sarah (2004). 'Vocal fighting: The functions of birdsong'. In P. Marler and H. W. Slabbekoorn (eds), *Nature's Music: The Science of Birdsong*. San Diego: Elsevier Academic Press, 39–79.

Conard, Nicholas J., Malina, Maria, and Münzel, Susanne C. (2009). 'New flutes document the earliest musical tradition in southwestern Germany'. *Nature* 460(7256): 737–740.

Coolidge, Frederick L., and Wynn, Thomas (2004). 'A cognitive and neuropsychological perspective on the Châtelperronian'. *Journal of Anthropological Research* 60: 55–73.

Coop, Graham, Bullaughey, Kevin, Luca, Francesca, and Przeworski, Molly (2008). 'The timing of selection at the human *FOXP2* gene'. *Molecular Biology and Evolution* 25: 1257–1259.

Copeland, B. Jack (2002). 'Hypercomputation'. *Minds and Machines* 12: 461–502.

Coqueugniot, Hélène, and Hublin, Jean-Jacques. (2008). 'Endocranial volume and brain growth in immature neanderthals'. *Periodicum Biologorum* 109(4): 379–385.

Corballis, Michael C. (2007). 'The uniqueness of human recursive thinking'. *American Scientist* 95(3): 240–248.

—— (2011). *The Recursive Mind. The Origins of Human Language, Thought, and Civilization*. Princeton, NJ: Princeton University Press.

Corbett, Greville G. (2006). *Agreement*. Cambridge: Cambridge University Press.

Cordemoy, Géraud de (1668). *Discurso físico de la palabra*. Málaga: Universidad de Málaga, translated by José Chamizo Domínguez, 1989 [*A Philosophical Discourse Concerning Speech; and A Discourse Written to a Learned Friar*. Ed. Barbara Ross. New York: Delmar, 1972].

Cosmides, Leda, and Tooby, John (1987). 'From evolution to behavior: Evolutionary Psychology as the missing link'. In J. Dupré (ed.), *The Latest on the Best. Essays on Evolution and Optimality*. Cambridge, MA: MIT Press, 277–306.

Costa, Daniel P., and Sinervo, Barry (2004). 'Field physiology: Physiological insight from animals in nature'. *Annual Review of Physiology* 66: 209–238.

Coyne, Jerry A., and Orr, H. Allen (2004). *Speciation*. Sunderland, MA: Sinauer.

Cummings, Jeffrey L. (1993). 'Frontal-subcortical circuits and human behavior'. *Archives of Neurology* 50: 873–880.

Curry, Haskell B. (1961). 'Some logical aspects of grammatical structure'. In R. Jakobson (ed.), *Structure of Language and its Mathematical Aspects. Proceedings of the Symposia in Applied Mathematics, Vol. XII*. Providence, RI: American Mathematical Society, 56–68.

d'Errico, Francesco (2003). 'The invisible frontier: A multispecies model for the origin of behavioral modernity'. *Evolutionary Anthropology* 12: 188–202.

—— Henshilwood, Christopher, Lawson, Graeme, Vanhaeren, Marian, Tillier, Anne-Marie, Soressi, Marie, Bresson, Frédérique, Maurielle, Bruno, Nowell, April, Lakarra, Joseba, Backwell, Lucinda, and Julien, Michèle (2003). 'Archaeological evidence for the emergence of language, symbolism, and music—An alternative multidisciplinary perspective'. *Journal of World Prehistory* 17(1): 1–70.

—— —— Vanhaeren, Marian, and van Niekerk, Karen (2005). '*Nassarius kraissianus* shell beads from Blombos Cave: evidence for symbolic behaviour in the Middle Stone Age'. *Journal of Human Evolution* 48: 3–24.

—— and Soressi, Marie (2002). 'Systematic use of manganese by the Pech-de-l'Azé Neandertals: Implications for the origin of behavioural modernity'. *Journal of Human Evolution* 42: A13.

Darwin, Charles (1859). *The Origin of Species*. London: John Murray.

—— (1871). *The Descent of Man, and Selection in Relation to Sex (Second edition, 1879)*. London: John Murray.

Davis, Martin (2004). 'The myth of hypercomputation'. In C. Teuscher (ed.), *Alan Turing: The Life and Legacy of a Great Thinker*. Berlin: Springer, 195–212.

Dawkins, Richard (1976). *The Selfish Gene*. Oxford: Oxford University Press.

—— (1982). *The Extended Phenotype. The Long Reach of the Gene*. Oxford: Oxford University Press.

—— and Krebs, John R. (1978). 'Animal signals: Information or manipulation?'. In J. R. Krebs and N. B. Davies (eds). *Behavioural Ecology: An Evolutionary Approach*. Oxford: Blackwell, 282–309.

de Beer, Gavin (1940). *Embryos and Ancestors*. Oxford: Clarendon Press.

—— (1971). *Homology, an Unsolved Problem*. Oxford: Oxford University Press.

de Queiroz, Alan, and Wimberger, Peter H. (1993). 'The usefulness of behavior for phylogeny estimation: Levels of homoplasy in behavioral and morphological characters'. *Evolution* 47: 46–60.

De Renzi, Miquel (2009). 'Developmental and historical patterns at the crossroads in the work of Pere Alberch'. In D. Rasskin-Gutman and M. De Renzi (eds), *Pere Alberch. The Creative Trajectory of an Evo-Devo Biologist*. Valencia: Institut d'Estudis Catalans and Universitat de València, 45–66.

—— Moya, Andrés, and Pereto, Juli (1999). 'Obituary. Evolution, development and complexity in Pere Alberch (1954–1998)'. *Journal of Evolutionary Biology* 12: 624–626.

Deacon, Terrence (1990a). 'Fallacies of progression in theories of brain-size evolution'. *International Journal of Primatology* 11: 193–236.

—— (1990b). 'Problems of ontogeny and phylogeny in brain-size evolution'. *International Journal of Primatology* 11: 237–282.

—— (2000). 'Heterochrony in brain evolution: Cellular versus morphological analyses'. In S. T. Parker, J. Langer, and L. McKinney (eds), *Biology, Brains, and Behavior: The Evolution of Human Development*. Santa Fe, NM: School of American Research Press, 41–88.

Dennett, Daniel C. (1978). *Brainstorms. Philosophical Essays on Mind and Psychology*. Cambridge, MA: MIT Press.

—— (1987). *The Intentional Stance*. Cambridge, MA: MIT Press.

—— (1991). *Consciousness Explained*. Boston: Little, Brown and Company.

—— (1995). *Darwin's Dangerous Idea. Evolution and the Meanings of Life*. New York: Simon & Schuster.

—— (1996). *Kinds of Minds. Toward an Understanding of Consciousness*. New York: Basic Books.

Descartes, René (1637). *Discourse on the Method*. In J. Cottingham, R. Stoothoff, and D. Murdoch (eds), *The Philosophical Writings of Descartes, Vol. I*. Cambridge: Cambridge University Press, 1985, 111–151.

—— (1648). 'Comments on a certain broadsheet'. In J. Cottingham, R. Stoothoff, and D. Murdoch (eds), *The Philosophical Writings of Descartes, Vol. I*. Cambridge: Cambridge University Press, 1985, 294–311.

Deutsch, Jean, and Le Guyader, Hervé (1998). 'The neuronal zootype. An hypothesis'. *Comptes Rendus de l'Académie des Sciences de Paris. Sciences de la Vie* 321: 713–719.

Dicicco-Bloom, Emanuel, Lu, Nairu, Pintar, John E., and Zhang, Jiwen (1998). 'The PACAP ligand/receptor system regulates cerebral cortical neurogenesis'. *Annals of the New York Academy of Sciences* 865: 274–289.

Dobzhansky, Theodosius (1937). *Genetics and the Origin of Species*. New York: Columbia University Press.

Dorus, Steve, Vallender, Eric J., Evans, Patrick D., Anderson, Jeffrey R., Gilbert, Sandra L., Mahowald, Michael, Wyckoff, Gerald J., Malcom, Christine M., and Lahn, Bruce T. (2004). 'Accelerated evolution of nervous system genes in the origin of *Homo sapiens*'. *Cell* 119: 1027–1040.

Dunbar, Robin I. M. (1996). *Grooming, Gossip, and the Evolution of Language*. Cambridge, MA: Harvard University Press.

Ebbesson, Sven O. E. (1980). 'The Parcellation Theory and its relation to interspecific variability in brain organization, evolutionary and ontogenetic development and neuronal plasticity'. *Cell and Tissue Research* 213: 179–212.

Ebeling, K. Smilla, and Spanier, Bonnie B. (2011). 'What made those penguins gay? Gender and sexuality politics in the zoo'. In J. A. Fisher (ed.). *Gender and the Science of Difference. Cultural Politics of Contemporary Science and Medicine*. Piscataway, NJ: Rutgers University Press, 126–146.

Echelard, Yann, Epstein, Douglas J., St-Jacques, Benoit, Shen, Liya, Mohler, Jym, McMahon, Jill A., and McMahon, Andrew P. (1993). 'Sonic hedgehog, a member of a family of putative signaling molecules, is implicated in the regulation of CNS polarity'. *Cell* 75: 1417–1430.

Eibl-Eibesfeldt, Iräneus (1984). *Die Biologie des menschlichen Verhaltens: Grundriss der Humanethologie*. Munich: Pieper [*Human Ethology*. New York: Aldine de Gruyter, 1989].

Embick, David, Marantz, Alec, Miyashita, Yasushi, O'Neil, Wayne, and Sakai, Kuniyoshi L. (2000). 'A syntactic specialization for Broca's area'. *Proceedings of the National Academy of Sciences USA* 97: 6150–6154.

Emery, Nathan J. (2006). 'Cognitive ornithology: The evolution of avian intelligence'. *Philosophical Transactions of the Royal Society B* 361(1465): 23–43.

Enard, Wolfgang, Przeworki, Molly, Fischer, Simon E., Lai, Cecilia S. L., Wiebe, Victor, Kitano, Takashi, Monaco, Anthony P., and Pääbo, Svante (2002). 'Molecular evolution of *FOXP2*, a gene involved in speech and language'. *Nature* 418: 868–872.

Etxeberria, Arantza, and Nuño de la Rosa, Laura (2009). 'A world of opportunity within constraint: Pere Alberch's early evo-devo'. In D. Rasskin-Gutman and M. De Renzi (eds), *Pere Alberch. The Creative Trajectory of an Evo-Devo Biologist*. Valencia: Institut d'Estudis Catalans and Universitat de València, 21–44.

Evans, Patrick D., Anderson, Jeffrey R., Vallender, Eric J., Choi, Sun Shim, and Lahn, Bruce T. (2004a). 'Reconstructing the evolutionary history of *microcephalin*, a gene controlling human brain size'. *Human Molecular Genetics* 13: 1139–1145.

——— Gilbert, Sandra L., Malcom, Christine. M., Dorus, Steve, and Lahn, Bruce T. (2004b). 'Adaptive evolution of *ASPM*, a major determinant of cerebral cortical size in humans'. *Human Molecular Genetics* 13: 489–494.

——— Mekel-Bobrov, Nitzan, Vallender, Eric J., Hudson, Richard R., and Lahn, Bruce T. (2006). 'Evidence that the adaptive allele of the brain size gene microcephalin introgressed into *Homo sapiens* from an archaic *Homo* lineage'. *Proceedings of the National Academy of Sciences of the United States of America* 103: 18178–18183.

Everett, Daniel (2005). 'Cultural constraints on grammar and cognition in Pirahã'. *Current Anthropology* 46(4): 621–46.

——— (2009). 'Pirahã culture and grammar: A response to some criticisms'. *Language* 85(2): 405–442.

Fadini, Ubaldo, Negri, Antonio, and Wolfe, Charles T. (eds) (2001). *Desiderio del mostro. Dal circo al laboratorio alla politica*. Rome: Manifestolibri.

Farris, Sarah M. (2008). 'Evolutionary convergence of higher brain centers spanning the protostome-deuterostome barrier'. *Brain, Behavior and Evolution* 72: 106–122.

Fay, Justin C., and Wu, Chung-I (1999). 'A human population bottleneck can account for the discordance between patterns of mitochondrial versus nuclear DNA variation'. *Molecular Biology and Evolution* 16(7): 1003–1005.

Ferland, Russell J., Cherry, Timothy J., Preware, Patricia O., Morrisey, Edward E., and Walsh, Christopher A. (2003). 'Characterization of Foxp2 and Foxp1 mRNA and protein in the developing and mature brain'. *Journal of Comparative Neurology* 460: 266-279.

Finlay, Barbara L. (2007). 'Endless minds most beautiful'. *Developmental Sciences* 10: 30-34.

—— and Darlington, Richard B. (1995). 'Linked regularities in the development and evolution of mammalian brains'. *Science* 268: 1578-1584.

Finlayson, Clive (2009). *The Humans Who Went Extinct. Why Neanderthals Died Out and We Survived.* New York: Oxford University Press.

Fitch, W. Tecumseh (2010). 'Three meanings of "recursion": Key distinctions for biolinguistics'. In R. K. Larson, V. Déprez, and H. Yamakido (eds), *The Evolution of Human Language. Biological Perspectives*. Cambridge: Cambridge University Press, 73-90.

—— and Hauser, Marc D. (2004). 'Computational constraints on syntactic processing in a nonhuman primate'. *Science* 303: 377-380.

—— —— Chomsky, Noam (2005). 'The evolution of the language faculty: Clarifications and implications'. *Cognition* 97: 179-210.

Fodor, Jerry A. (1968). *Psychological Explanation*. New York: Random House.

—— (1975). *The Language of Thought*. New York: Thomas Crowell.

—— (1981). *RePresentations*. Cambridge, MA: MIT Press.

—— (1983). *The Modularity of Mind*. Cambridge, MA: MIT Press.

—— (1986). 'Why paramecia don't have mental representations'. *Midwest Studies in Philosophy* X: 3-23.

—— (2000). *The Mind Doesn't Work that Way. The Scope and Limits of Computational Psychology*. Cambridge, MA: MIT Press.

—— (2008). *LOT 2. The Language of Thought Revisited*. Oxford: Oxford University Press.

—— Piattelli-Palmarini, Massimo (2010). *What Darwin Got Wrong*. New York: Farrar, Straus & Giroux.

Fortnow, Lance, and Homer, Steve (2003). 'A short history of computational complexity'. *Bulletin of the European Association for Theoretical Computer Science* 80: 95-133.

Foucault, Michel (1961). *Folie et déraison. Histoire de la folie à l'âge classique*. Paris: Librairie Plon.

—— (1999). *Les anormaux. Cours au Collège de France. 1974-1975*. Paris: Gallimard/Le Seuil.

Friederici, Angela D. (2002). 'Towards a neural basis for auditory sentence processing'. *Trends in Cognitive Sciences* 6: 78-84.

Frith, Clifford B., Borgia, Gerald, and Frith, Dawn W. (1996). 'Courts and courtship behaviour of Archbold's Bowerbird *Archboldia papuensis* in Papua New Guinea'. *Ibis* 138: 204-211.

Futuyma, Douglas J. (2005). *Evolution*. Sunderland, MA: Sinauer.

Gallistel, C. Randy, and King, Adam P. (2009). *Memory and the Computational Brain. Why Cognitive Science Will Transform Neuroscience*. New York: Blackwell/Wiley.

Gao, Yijie, Sun, Yi, Frank, Karen M., Dikkes, Pieter, Fujiwara, Yuko: Seidl, Katherine J., Sekiguchi, JoAnn M., Rathbun, Gary A., Swat, Wojciech, Wang, Jiyang, Bronson, Roderick T., Malynn, Barbara A., Bryans, Margaret, Zhu, Chengming, Chaudhuri, Jayanta, Davidson, Laurie, Ferrini, Roger, Stamato, Thomas, Orkin, Stuart H., Greenberg, Michael E., and Alt, Frederick W. (1998). 'A critical role for DNA end-joining proteins in both lymphogenesis and neurogenesis'. *Cell* 95: 891-902.

García-Azkonobieta, Tomás (2005). *Evolución, desarrollo y (auto) organización. Un estudio sobre los principios filosóficos de la Evo-Devo*. Donostia: Universidad del País Vasco dissertation.

Garland, Theodore Jr., and Kelly, Scott A. (2006). 'Phenotypic plasticity and experimental evolution'. *The Journal of Experimental Biology* 209: 2344–2361.

Gathorne-Hardy, F. J., and Harcourt-Smith, W. E. H. (2003). 'The super-eruption of Toba, did it cause a human bottleneck?'. *Journal of Human Evolution* 45: 227–230.

Gazdar, Gerald, Klein, Ewan, Pullum, Geoffrey, and Sag, Ivan (1985). *Generalized Phrase Structure Grammar*. Oxford: Basil Blackwell.

Geissmann, Thomas (2000). 'Gibbon songs and human music from an evolutionary perspective'. In N. L. Wallin, B. Merker, and S. Brown (eds). *The Origins of Music*. Cambridge, MA: MIT Press, 103–123.

Gentner, Timothy Q., Fenn, Kimberly M., Margoliash, Daniel, and Nusbaum, Howard C. (2006). 'Recursive syntactic pattern learning by songbirds'. *Nature* 440: 1204–1207.

Geoffroy Saint-Hilaire, Étienne (1818). *Philosophie anatomique*. Paris: J.-B. Baillière.

—— (1820). *Mémoires sur l'organisation des insectes*. Paris: Pancoucke.

—— (1830). *Principes de philosophie zoologique*. Paris: Pichon et Didier.

Geoffroy Saint-Hilaire, lsidore (1832). *Histoire générale et particulière des anomalies de l'organisation chez l'homme et les animaux ou Traité de tératologie, Vol I*. Paris: J.-B. Baillière.

—— (1837). *Histoire générale et particulière des anomalies de l'organisation chez l'homme et les animaux ou Traité de tératologie. Atlas contenant 20 planches avec leur explication, et table générale des matières*. Paris: J.-B. Ballière.

Ghiselin, Michael T. (2005). 'Homology as a relation of correspondence between parts of individuals'. *Theory in Biosciences* 124: 91–103.

Gibson, Robert M., Bradbury, Jack W., and Vehrencamp, Sandra L. (1991). 'Mate choices in lekking sage grouse revisited: The roles of vocal display, female site fidelity and copying'. *Behavioral Ecology* 2: 162–180.

Gilbert, Scott F. (2003a). 'The morphogenesis of evolutionary developmental biology'. *International Journal of Developmental Biology* 47: 467–477.

—— (2003b). *Developmental Biology*. Eighth edition. Sunderland, MA: Sinauer.

—— (2003c). 'Evo-Devo, Devo-Evo, and Devgen-Popgen'. *Biology and Philosophy* 18: 347–352.

—— Bolker, Jessica A. (2001). 'Homologies of process and molecular elements of embryonic construction'. *Journal of Experimental Zoology (Mol Dev Evol)* 291: 1–12.

—— Opitz, John M., and Raff, Rudolf A. (1996). 'Resynthesizing evolutionary and developmental biology'. *Developmental Biology* 173: 357–372.

Godfrey-Smith, Peter (1994). 'A modern theory of functions'. *Noûs* 28: 344–362.

Goethe, Johann Wolfgang von (1795). 'Erster Entwurf einer allgemeinen Einleitung in die vergleichende Anatomie, ausgehend von der Osteologie'. In *Zur Morphologie, Vol. 1*, Book 2, Stuttgart, 1820. Available online at <http://www.merke.ch/goethe/anatomie/erster_entwurf.php>.

Golani, Ilan (1992). 'A mobility gradient in the organization of vertebrate movement: The perception of movement through symbolic language'. *Behavioral and Brain Sciences* 15: 249–308.

Goldschmidt, Richard (1940). *The Material Basis of Evolution*. New Haven, CT: Yale University Press.
Goodman, Corey S., and Coughlin, Bridget C. (2000). 'Introduction. The evolution of evo-devo biology'. *Proceedings of the National Academy of Science* 978(9): 4424–4425.
Goodman, Nelson (1954). *Fact, Fiction, and Forecast*. Cambridge, MA: Harvard University Press.
Goodwin, Brian (1994). *How the Leopard Changed Its Spots: The Evolution of Complexity*. London: Phoenix.
Gould, Stephen J. (1977). *Ontogeny and Phylogeny*. Cambridge, MA: Harvard University Press.
—— (1989). *Wonderful Life. The Burgess Shale and the Nature of History*. New York: W.W. Norton & Company.
—— (2002). *The Structure of Evolutionary Theory*. Cambridge, MA: Belknap Press of Harvard University Press.
—— and Lewontin, Richard (1979). 'The spandrels of San Marco and the Panglossian paradigm: A critique of the adaptationist program'. *Proceedings of the Royal Society B* 205: 581–589.
Gräff, Johannes, and Mansuy, Isabelle M. (2008). 'Epigenetic codes in cognition and behavior'. *Behavioral Brain Research* 192: 70–87.
Grahn, Jessica A. (2009). 'The role of the basal ganglia in beat perception'. *Annals of the New York Academy of Sciences* 1169: 35–45.
Grassé, Pierre-Paul (1959). 'La reconstruction du nid et les coordinations inter-individuelles chez *Bellicostermes natalensis* et *Cubitermes* sp. La théorie de la stigmergie: essai d'interprétation du comportement des termites constructeurs'. *Insectes Sociaux* 6: 41–81.
Greene, Harry W. (1994). 'Homology and behavioral repertoires'. In B. K. Hall (ed.), *Homology. The Hierarchical Basis of Comparative Biology*. San Diego: Academic Press, 369–391.
—— (1999). 'Natural history and behavioural homology'. In G. R. Bock and G. Cardew (eds), *Homology*. Chichester: John Wiley & Sons/Novartis Foundation, 173–188.
Griesser, Michael (2008). 'Referential calls signal predator behavior in a group-living bird species'. *Current Biology* 18: 69–73.
Griffiths, Paul E. (1994). 'Cladistic classification and functional explanation'. *Philosophy of Science* 61: 206–227.
—— (1997). *What Emotions Really Are. The Problem of Psychological Categories*. Chicago: University of Chicago Press.
—— (2004a). 'Is emotion a natural kind?' In R. C. Solomon (ed.), *Thinking about Feeling: Contemporary Philosophers on Emotions*. Oxford: Oxford University Press, 233–249
—— (2004b). 'Emotions as natural and normative kinds'. *Philosophy of Science* 71: 901–911
—— (2007). 'Evo-Devo meets the mind: Towards a developmental evolutionary psychology'. In R. Brandon and R. Sansom (eds), *Integrating Evolution and Development: From Theory to Practice*. Cambridge, MA: MIT Press: 195–226.
—— and Gray, Russell D. (1994). 'Developmental systems and evolutionary explanations'. *The Journal of Philosophy* XCI: 277–304.
—— and Stotz, Karola (2000). 'How the mind works: A developmental perspective on the biology of cognition'. *Synthese* 122: 29–51.

Grodzinsky, Yosef (2000). 'The neurology of syntax: Language use without Broca's area'. *Behavioral and Brain Sciences* 23: 1–71.

Grün, Rainer, and Stringer, Chris (2000). 'Tabun revisited: Revised ESR chronology and new ESR and U-series analyses of dental material from Tabun C1'. *Journal of Human Evolution* 39: 601–612.

Gunz, Philipp, Neubauer, Simon, Golovanova, Lubov, Doronichev, Vladimir, Maureille, Bruno, and Hublin, Jean-Jacques. (2012). 'A uniquely modern human pattern of endocranial development. Insights from a new cranial reconstruction of the Neandertal newborn from Mezmaiskaya'. *Journal of Human Evolution*, doi: 10.1016/j.jhevol.2011.11.013.

—— Maureille, Bruno, and Hublin, Jean-Jacques (2010). 'Brain development after birth differs between Neanderthals and modern humans'. *Current Biology* 20(21): R921–R922.

Haesler, Sebastian, Rochefort, Christelle, Georgi, Benjamin, Licznerki, Pawel, Osten, Pavel, and Scharff, Constance (2007). 'Incomplete and inaccurate vocal imitation after knockdown of FoxP2 in songbird basal ganglia nucleus Area X'. *PLoS Biology* 5(12): e321.

—— Wada, Kazuhiro, Nshdejan, A., Morrisey, Edward, Lints, Thierry, Jarvis, Eric D., and Scharff, Constance (2004). 'FOXP2 expression in avian vocal learners and non-learners'. *The Journal of Neurosciences* 24 (13): 3164–3175.

Hakem, Razqallah, de la Pompa, José Luis, Sirard, Christian, Mo, Rong, Woo, Minna, Hakem, Anne, Wakeham, Andrew, Potter, Julia, Reitmair, Armin, Billia, Filio, Firpo, Eduardo, Hui, Chi Chung, Roberts, Jim, Rossant, Janet, and Mak, Tak W. (1996). 'The tumor suppressor gene *Brca1* is required for embryonic cellular proliferation in the mouse'. *Cell* 85: 1009–1023.

Hall, Brian K. (1999). *Evolutionary Developmental Biology*. Second edition. Dordrecht: Kluwer.

—— (2001). 'Evolutionary developmental biology: Where embryos and fossils meet'. In N. Minugh-Purvis and K. J. McNamara (eds), *Human Evolution through Developmental Change*. Baltimore, MD: Johns Hopkins University Press, 7–27.

—— (2003). 'Evo-Devo: Evolutionary developmental mechanisms'. *International Journal of Developmental Biology* 47: 491–495.

—— (ed.) (1994). *Homology: The Hierarchical Basis of Comparative Biology*. San Diego: Academic Press.

—— and Olson, Wendy M. (eds) (2003a). *Keywords and Concepts in Evolutionary Developmental Biology*. Cambridge, MA: Harvard University Press.

—— —— (eds) (2003b). 'Introduction'. In B. K. Hall and W. M. Olson (eds), *Keywords and Concepts in Evolutionary Developmental Biology*. Cambridge, MA: Harvard University Press, xiii–xvi.

Hamburger, Viktor (1980). 'Embriology and the Modern Synthesis'. In E. Mayr and W. B. Provine (eds), *The Evolutionary Synthesis: Perspectives on the Unification of Biology*. Cambridge, MA: Harvard University Press: 97–112.

Hansell, Mike H. (2000). *Bird Nests and Construction Behaviour*. Cambridge: Cambridge University Press.

—— (2005). *Animal Architecture*. Oxford: Oxford University Press.

Harpending, Henry C., Sherry, Stephen T., Rogers, Alan R., and Stoneking, Mark (1993). 'The genetic structure of ancient human populations'. *Current Anthropology* 34(4): 483–496.

Harzsch, Steffen (2006). 'Neurophylogeny: Architecture of the nervous system and a fresh view on arthropod phylogeny'. *Integrative and Comparative Biology* 46(2): 162–194.

Hass, Joel, Lagarias, Jeffrey C., and Pippenger, Nicholas (1999). 'The computational complexity of knots and link problems'. *Journal of the ACM* 46(2): 185–211.

Hauser, Marc D. (1997). *The Evolution of Communication*. Cambridge, MA: MIT Press.

—— Chomsky, Noam, and Fitch, W. Tecumseh (2002). 'The faculty of language: What is it, who has it, and how did it evolve?'. *Science* 298: 1569–1579.

Hawks, John, Hunley, Keith, Lee, Sang-Hee, and Wolpoff, Milford (2000). 'Population bottlenecks and Pleistocene human evolution'. *Molecular Biology and Evolution* 17: 2–22.

Heine, Bernd, and Kuteva, Tania (2007). *The Genesis of Grammar. A Reconstruction.* Oxford: Oxford University Press.

Hendrikse, Jesse L., Parsons, Trish E., and Hallgrímsson, Benedikt (2007). 'Evolvability as the proper focus of evolutionary developmental biology'. *Evolution and Development* 9: 393–401.

Henning, Willi (1966). *Phylogenetic Systematics*. Urbana, IL: University of Illinois Press.

Henshilwood, Christopher S., d'Errico, Franceso, Yates, Royden, Jacobs, Zenobia, Tribolo, Chantal, Duller, Geoff A. T., Mercier, Norbert, Sealy, Judith C., Valladas, Helene, Watts, Ian, and Wintle, Ann G. (2002). 'Emergence of modern human behaviour: Middle Stone Age engravings from South Africa'. *Science* 295: 1278–1280.

Herder, Johann Gottfried von (1772). 'Ensayo sobre el origen del lenguaje'. In *Obra selecta*. Madrid: Alfaguara, translated by Pedro Ribas, 1982, 131–232 [*Treatise on the Origin of Language*. In *Herder. Philosophical Writings*. Ed. Michael N. Foster. Cambridge: Cambridge University Press, 2002].

Herzfeld, Chris, and Lestel, Dominique (2005). 'Knot tying in great apes: Etho-ethnology of an unusual tool behavior'. *Social Science Information* 44(4): 621–653.

Hetherington, Renée, and Reid, Robert G. B. (2010). *The Climate Connection. Climate Change and Modern Human Evolution.* Cambridge: Cambridge University Press.

Hinde, Robert A. (1958). 'The nest building behaviour of domesticated canaries'. *Proceedings of the Zoological Society of London* 131: 1–48.

Ho, Yu-Chi, and Pepyne, David L. (2002). 'Simple explanation of the No-Free Lunch Theorem and its implications'. *Journal of Optimization Theory and Applications* 115(3): 549–570.

Hobbes, Thomas (1641). 'Third set of objections to Descartes' Meditations'. In J. Cottingham, R. Stoothoff, and D. Murdoch (eds), *The Philosophical Writings of Descartes, Vol. I*. Cambridge: Cambridge University Press, 1985, 121–137.

Hockett, Charles F. (1958). *A course in modern linguistics*. New York: The Macmillan Company.

—— (1960). 'The origin of speech'. *Scientific American* 203: 89–97.

Hodos, William (1976). 'The concept of homology and the evolution of behavior'. In R. B. Masterton, W. Hodos, and H. Jerison (eds), *Evolution, Brain, and Behavior: Persistent Problems*. Hillsdale, NJ: Lawrence Erlbaum, 153–167.

Hofman, Michel A. (2001). 'Brain evolution in hominids: Are we at the end of the road?'. In D. Falk and K. R. Gibson (eds), *Evolutionary Anatomy of the Primate Cerebral Cortex*. Cambridge: Cambridge University Press, 113–127.

Hopcroft, John E., and Ullman, Jeffrey D. (1979). *Introduction to Automata Theory, Languages, and Computation*. Reading, MA: Addison-Wesley.

Hornstein, Norbert (2009). *A Theory of Syntax. Minimal Operations and Universal Grammar.* Cambridge: Cambridge University Press.

Hung, Liang-Yu, Chen, Huang-Ling, Chang, Ching-Wen, Li, Bor-Ran, and Tang, Tang K. (2004). 'Identification of a novel microtubule-destabilizing motif in CPAP that binds to tubulin heterodimers and inhibits microtubule assembly'. *Molecular Biology of the Cell* 15: 2697–2706.
——Tang, Chieh-Ju C., and Tang, Tang K. (2000). 'Protein 4.1 R-135 interacts with a novel centrosomal protein (CPAP) which is associated with the gamma-tubulin complex'. *Molecular and Cellular Biology* 20: 7813–7825.
Hunley, Keith, Bowern, Claire, and Healy, Meghan (2012). 'Rejection of a serial founder effects model of genetic and linguistic coevolution'. *Proceedings of the Royal Society B*, doi: 10.1098/rspb.2011.2296.
Hunt, Gavin R. (1996). 'Manufacture and tool use of hook-tools by New Caledonian crows'. *Nature* 379: 249–251.
——and Gray, Russell D. (2003). 'Diversification and cumulative evolution in New Caledonian crow tool manufacture'. *Proceedings of the Royal Society of London B* 270(1517): 867–874.
Huxley, Julian (1932). *Problems of Relative Growth*. London: Methuen and Company.
——(1942). *Evolution. The Modern Synthesis*. London: Allen and Unwin.
Iwaniuk, Andrew N., Lefebvre, Louis, and Wylie, Douglas R. (2009). 'The comparative approach and brain-behaviour relationships: A tool for understanding tool use'. *Canadian Journal of Experimental Psychology/Revue Canadienne de Psychologie Expérimentale* 63(2): 150–159.
Jablonka, Eva, and Lamb, Marion J. (2005). *Evolution in Four Dimensions: Genetic, Epigenetic, Behavioral, and Symbolic Variation in the History of Life*. Cambridge, MA: MIT Press.
Jackendoff, Ray (2002). *Foundations of Language: Brain, Meaning, Grammar, Evolution*. Oxford: Oxford University Press.
——and Pinker, Steven (2005). 'The nature of the language faculty and its implications for evolution of language (Reply to Fitch, Hauser, and Chomsky)'. *Cognition* 97: 211–225.
Jenkins, Lyle (2000). *Biolinguistics. Exploring the Biology of Language*. Cambridge: Cambridge University Press.
Jespersen, Otto (1922). *Language. Its Nature, Development and Origin*. London: George Allen & Unwin Ltd.
Johnston, Timothy D., and Edwards, Laura (2002). 'Genes, interactions, and the development of behaviour'. *Psychological Review* 109: 26–34.
Jones, Clive G., Lawton, John H., and Shachak, Moshe (1994). 'Organisms as ecosystems engineers'. *Oikos* 69: 373–386.
——————(1997). 'Positive and negative effects of organisms as physical ecosystem engineers'. *Ecology* 78: 1946–1957.
Joshi, Aravind K. (1985). 'Tree adjoining grammars: How much context-sensitivity is required to provide reasonable structural descriptions?'. In D. R. Dowty, L. Karttunen, and A. M. Zwicky (eds), *Natural Language Parsing: Psychological, Computational, and Theoretical Perspectives*. Cambridge: Cambridge University Press, 206–250.
——and Schabes, Yves (1997). 'Tree Adjoining Grammars'. In G. Rozenberg and A. Salomaa (eds), *Handbook of Formal Languages. Vol. 3*. Berlin: Springer, 69–126.

—— Vijay-Shanker, K., and Weir, David (1991). 'The converge of mildly context-sensitive grammar formalisms'. In P. Sells, S. M. Shieber, and T. Wasow (eds), *Foundational Issues in Natural Language Processing*. Cambridge, MA: MIT Press, 31–81.

Karten, Harvey J., and Shimizu, Tom (1989). 'The origins of neocortex: Connections and lamination as distinct events in evolution'. *Journal of Cognitive Neuroscience* 1: 291–301.

Kaskan, Peter M., and Finlay, Barbara L. (2001). 'Encephalization and its developmental structure: How many ways can a brain get big?'. In D. Falk and K. R. Gibson (eds), *Evolutionary Anatomy of the Primate Cerebral Cortex*. Cambridge: Cambridge University Press, 14–19.

Kauffman, Stuart (1993). *The Origins of Order. Self-Organization and Selection in Evolution*. New York: Oxford University Press.

—— (1995). *At Home in the Universe: The Search for Laws of Complexity*. New York: Oxford University Press.

Keller, Evelyn Fox (2000). *The Century of the Genes*. Cambridge, MA: Harvard University Press.

Kelso, J. A. Scott (1995). *Dynamic Patterns. The Self-Organization of Brain and Behavior*. Cambridge, MA: MIT Press.

Kennedy, John S. (1992). *The New Anthropomorphism*. Cambridge: Cambridge University Press.

Ketterson, Ellen D., and Nolan Jr., Val. (1999). 'Adaptation, exaptation, and constraint: A hormonal perspective'. *The American Naturalist* 154(S1): S4–S25.

Khabbaz, Nabil A. (1974). 'A geometric hierarchy of languages'. *Journal of Computer and System Sciences* 8: 142–157.

Kiparsky, Paul (2008). 'Universals constrain change, change results in typological generalizations'. In J. Good (ed.), *Linguistic Universals and Language Change*. Oxford: Oxford University Press, 23–53.

—— (2011). 'Grammaticalization as optimization'. In D. Jonas, J. Whitman, and A. Garrett (eds), *Grammatical Change: Origins, Nature, Outcomes*. Oxford: Oxford University Press, 15–51.

Kirby, Simon (2002). 'Learning bottlenecks and the evolution of recursive syntax'. In T. Briscoe (ed.), *Linguistic Evolution through Language Acquisition: Formal and Computational Models*. Cambridge: Cambridge University Press, 173–204.

—— Dowman, Mike, and Griffiths, Thomas L. (2007). 'Innateness and culture in the evolution of language'. *Proceedings of the National Academy of Sciences USA* 104: 5241–5245.

Klein, Richard G. (2009). *The Human Career. Human Biological and Cultural Origins*. Third edition. Chicago: University of Chicago Press.

—— Edgar, Blake (2002). *The Dawn of Human Culture*. New York: John Wiley.

Klingberg, Torkel (2009). *The Overflowing Brain: Information Overload and the Limits of Working Memory*. Oxford: Oxford University Press.

Klopfer, Peter H. (1969). 'Review of R. F. Ewer, *Ethology of Mammals*'. *Science* 165: 887.

—— (1973a). 'Does behavior evolve?'. *Annals of the New York Academy of Sciences* 223: 113–125.

—— (1973b). 'Evolution and behavior'. In G. Bermant (ed.), *Perspectives on Animal Behavior*. Chicago: Scott, Foreman and Co., 48–71.

Klopfer, Peter H. (1974). *An Introduction to Animal Behavior. Ethology's First Century*. Second edition. Englewood Cliffs, NJ: Prentice Hall.
Kofron, Christopher P. (1993). 'Behavior of the Nile crocodiles in a seasonal river in Zimbabwe'. *Copeia* 2: 463–469.
Krause, Johannes, Lalueza-Fox, Carles, Orlando, Ludovic, Enard, Wolfgang, Green, Richard E., Burbano, Hernán A., Hublin, Jean-Jacques, Hänni, Catherine, Fortea, Javier, de la Rasilla, Marco, Bertranpetit, Jaume, Rosas, Antonio, and Pääbo, Svante (2007). 'The derived *FOXP2* variant of modern humans was shared with Neandertals'. *Current Biology* 17: 1908–1912.
Krebs, John R., and Davies, Nick B. (1987). *An Introduction to Behavioural Ecology*. Second edition. Oxford: Blackwell.
—— and Dawkins, Richard (1984). 'Animal signals: Mind reading and manipulation'. In J. R. Krebs and N. B. Davies (eds), *Behavioral Ecology: An Evolutionary Approach*. Second edition. Sunderland, MA: Sinauer: 380–402.
Krink, Thiemo, and Vollrath, Fritz (1998). 'Emergent properties in the behaviour of a virtual spider robot'. *Proceedings of the Royal Society of London B* 265: 2051–2055.
—— —— (1999). 'A virtual robot to model the use of regenerated legs in a web-building spider'. *Animal Behaviour* 57: 223–232.
Kuida, Keisuke, Zheng, Timothy S., Na, Songqing, Kuan, Chia-Yi, Yang, Di, Karasuyama, Hajime, Rakic, Pasko, and Flavell, Richard A. (1996). 'Decreased apoptosis in the brain and premature lethality in CPP32-deficient mice'. *Nature* 384: 368–372.
Kuno, Susumo (1973). *The Structure of Japanese Language*. Cambridge, MA: MIT Press.
La Mettrie, Julien Offray de (1747). *El hombre máquina. El arte de gozar*. Madrid: Valdemar, translated by Agustín Izquierdo and María Badiola, 2000 [*Machine Man and Other Writings*. Ed. Ann Thomson. Cambridge: Cambridge University Press, 1996].
Labov, William A. (1994). *Principles of Linguistic Change. Vol. 1: Internal Factors*. Oxford: Blackwell.
—— (2001). *Principles of Linguistic Change. Vol. 2: Social Factors*. Oxford: Blackwell.
—— (2010). *Principles of Linguistic Change. Vol. 3. Cognitive and Cultural Factors*. Malden, MA: Wiley-Blackwell.
Lahr, M. M., and Foley, R. A. (1998). 'Towards a theory of human origins: Geography, demography and diversity in recent human evolution'. *Yearbook of Physical Anthropology* 41: 137–176.
Lai, Cecilia S., Fisher, Simon E., Hurst, Jane A., Vargha-Khadem, Faraneh, and Monaco, Anthony P. (2001). 'A forkhead-domain gene is mutated in a severe speech and language disorder'. *Nature* 413: 519–523.
—— Gerrelli, Dianne, Monaco, Anthony P., Fisher, Simon E., and Copp, Andrew J. (2003). '*FOXP2* expression during brain development coincides with adult sites of pathology in a severe speech and language disorder'. *Brain* 126: 2455–2462.
Langer, Jonas (2000). 'The heterochronic evolution of primate cognitive development'. In S. T. Parker, J. Langer, and L. McKinney (eds), *Biology, Brains, and Behavior: The Evolution of Human Development*. Santa Fe, NM: School of American Research Press, 215–235.
—— (2006). 'The heterochronic evolution of primate cognitive development'. *Biological Theory* 1(1): 41–43.

Laubichler, Manfred D. (2010). 'Evolutionary developmental biology offers a significant challenge to the Neo-Darwinian paradigm'. In F. J. Ayala and R. Arp (eds), *Contemporary Debates in Philosophy of Biology*. Chichester: Wiley-Blackwell, 199-212.

—— and Maienschein, Jane (eds) (2007). *From Embryology to Evo-Devo: A History of Developmental Evolution*. Cambridge, MA: MIT Press.

Lauder, George V. (1994). 'Homology, form, and function'. In B. K. Hall (ed.), *Homology: The Hierarchical Basis of Comparative Biology*. San Diego: Academic Press:151-196.

Lefebvre, Louis, Nicolakakis, Nektaria, and Boire, Denis (2002). 'Tools and brains in birds'. *Behaviour* 139: 939-973.

Leibniz, Gottfried Wilhelm (1765). *Nuevos ensayos sobre el entendimiento humano*. Madrid: Alianza, translated by J. Echeverría Ezponda, 1992 [*New Essays on Human Understanding*. Ed. Peter Remnant and Jonathan Bennett. Cambridge: Cambridge University Press, 1996].

Lenneberg, Eric (1967). *Biological Foundations of Language*. New York: John Wiley and Sons, Inc.

Lewis, Harry R., and Papadimitriou, Christos H. (1981). *Elements of the Theory of Computation*. Englewood Cliffs, NJ: Prentice Hall.

Lewis-Williams, David (2002). *The Mind in the Cave. Conciousness and the Origins of Art*. London: Thames and Hudson.

Lewontin, Richard C. (1983). 'Gene, organism and environment'. In D. S. Bendall (ed.), *Evolution from Molecules to Men*. Cambridge: Cambridge University Press, 273-285.

—— (1998). 'The evolution of cognition: Questions we will never answer'. In D. Scarborough and S. Sternberg (eds), *An Invitation to Cognitive Science. Methods, Models, and Conceptual Issues*. Vol. 4. Second edition. Cambridge, MA: MIT Press, 107-132.

—— Rose, Steven, and Kamin, Leon J. (1984). *Not in Our Genes. Biology, Ideology, and Human Nature*. New York: Pantheon.

Lieberman, Philip (2006). *Toward an Evolutionary Biology of Language*. Cambridge, MA: Harvard University Press.

Liégeois, Frédérique, Badelweg, Torsten, Connelly, Alan, Gadian, David G., Mishkin, Mortimer, and Vargha-Khadem, Faraneh (2003). 'Language fMRI abnormalities associated with *FOXP2* gene mutation'. *Nature Neuroscience* 6: 1230-1237.

Linde Medina, Marta (2010). 'Two "EvoDevos"'. *Biological Theory* 5(1): 7-11.

Lorenz, Konrad (1974). 'Analogy as a source of knowledge'. *Science* 185: 229-234.

Lorenzo, Guillermo (2008). 'Los límites de la selección natural y el evo-minimalismo. Antecedentes, actualidad y perspectivas del pensamiento chomskyano sobre los orígenes evolutivos del lenguaje'. *Verba* 35: 387-421.

—— (2010). 'Devo-Darwinismo. Lo que el lenguaje nos enseña sobre el papel del desarrollo en la evolución natural'. *Éndoxa* 24 (monographic issue on Darwin): 247-274.

—— and Longa, Víctor M. (2011). 'Developmental plasticity and linguistic variation'. Ms., Universidad de Oviedo and Universidad de Santiago de Compostela.

Love, Alan C. (2007). 'Functional homology and homology of function: Biological concepts and philosophical consequences'. *Biology and Philosophy* 22: 691-708.

Lucretius (2003). *La naturaleza*. Madrid: Gredos, translated by Francisco Socas [*The Nature of Things*. Ed. Alicia E. Stallings. London: Penguin, 2007].

Madden, Joah (2001). 'Sex, bowers and brains'. *Proceedings of the Royal Society of London B* 268 (1469): 833–838.

Marigo, Valeria, Roberts, Drucilla J., Lee, Scott M. K., Tsukurov, Olga, Levi, Tatjana, Gastier, Julie M., Epstein, Debra J., Gilbert, Debra J., Copeland, Neal G., Seidman, Christine E., Jenkins, Nancy A., Seidman, J. G., McMahon, Andrew P., and Tabin, Cliff (1995). 'Cloning, expression, and chromosomal location of SHH and IHH: two human homologues of the Drosophila segment polarity gene hedgehog'. *Genomics* 28: 44–51.

Marler, Peter (1977). 'The structure of animal communication sounds'. In T. H. Bullock (ed.), *Recognition of Complex Acoustic Signals (Report of the Dahlem Workshop)*. Berlin: Abakon-Verlagsgesellschaft, 17–35.

—— (1998). 'Animal communication and human language'. In N. G. Jablonski and L. C. Aiello (eds), *The Origin and Diversification of Language*. San Francisco: Memoirs of the California Academy of Sciences 24, 1–19.

—— and Slabbekoorn, Hans Willem (eds) (2004). *Nature's Music. The Science of Birdsong*. San Diego: Elsevier Academic Press.

Marr, David (1982). *Vision*. San Francisco: Freeman.

Marx, Karl, and Engels, Friedrich (1845/46). *La ideologia alemanya* (2 vols.) Barcelona: Laia, translated by Jordi Moners i Sinyol, 1987 [*The German Ideology*. Amherst, NY: Prometheus Books, 1998].

Mattick, John S. (2007). 'A new paradigm for developmental biology'. *The Journal of Experimental Biology* 210: 1526–1547.

—— Taft, Ryan J., and Faulkner, Geoffrey J. (2009). 'A global view of genomic information—Moving beyond the gene and the master regulator'. *Trends in Genetics* 26(1): 21–28.

Maynard Smith, John (1974). 'The theory of games and the evolution of animal conflicts'. *Journal of Theoretical Biology* 47: 209–221.

—— Burian, Richard, Kauffman, Stuart, Alberch, Pere, Campbell, J., Goodwin, Brian, Lande, Russell, Raup, David, and Wolpert, Lewis (1985). 'Developmental constraints and evolution: A perspective from the Mountain Lake conference on development and evolution'. *The Quarterly Review of Biology* 60: 265–287.

—— and Parker, Geoffrey A. (1976). 'The logic of asymmetric contests'. *Animal Behaviour* 24: 159–175.

—— and Price, George R. (1973). 'The logic of animal conflict'. *Nature* 246: 15–18.

Mayr, Ernst (1942). *Systematics and the Origin of Species*. Cambridge, MA: Harvard University Press.

—— (1963). *Animal Species and Evolution*. Cambridge, MA: Harvard University Press.

—— (1982). *The Growth of Biological Thought: Diversity, evolution, and inheritance*. Cambridge, MA: Belknap Press.

—— and Provine, William B. (eds) (1980). *The Evolutionary Synthesis: Perspectives on the Unification of Biology*. Cambridge, MA: Harvard University Press.

McBrearty, Sally, and Brooks, Alison S. (2000). 'The revolution that wasn't: A new interpretation of the origin of modern human behavior'. *Journal of Human Evolution* 39: 453–563.

McKaye, Kenneth R. (1979). 'Sexual selection and the evolution of cichlid fishes of Lake Malawi'. In M. H. A. Keenleyside (ed.), *Cichlid Fishes: Behaviour, Ecology and Evolution*. London: Chapman and Hall, 241–257.

McKinney, Michael L. (2000). 'Evolving behavioral complexity by extending development'. In S. T. Parker, J. Langer, and M. L. McKinney (eds), *Biology, Brains, and Behavior: The Evolution of Human Development*. Santa Fe, NM: School of American Research Press, 25-40.
—— and McNamara, Kenneth (1991). *Heterochrony: The Evolution of Ontogeny*. New York: Plenum.
Mehler, Mark F., and Mattick, John S. (2007). 'Noncoding RNAs and RNA editing in brain development, functional diversification, and neurological disease'. *Physiological Review* 87: 799-823.
Mekel-Bobrov, Nitzan, Gilbert, Sandra L., Evans, Patrick D., Vallender, Eric J., Anderson, Jeffrey R., Hudson, Richard R., Tishkoff, Sarah A., and Lahn, Bruce T. (2005). 'Ongoing adaptive evolution of *ASPM*, a brain size determinant in *Homo sapiens*'. *Science* 309: 1720-1722.
Mellars, Paul (1996). *The Neanderthal Legacy: An Archaeological Perspective*. Princeton, NJ: Princeton University Press.
—— (1998). 'Neanderthals, modern humans and the archaeological record'. In N. Jablonski and L. C. Aiello (eds), *The Origin and Diversification of Language*. San Francisco: Memoirs of the California Academy of Sciences 24, 89-115.
—— (2005). 'The impossible coincidence. A single-species model for the origins of modern human behavior'. *Evolutionary Anthropology* 14: 12-27.
Mellars, Paul, and Stringer, Chris (1989). *The Human Revolution: Behavioral and Biological Perspectives on the Origins of Modern Humans*. Edinburgh: Edinburgh University Press.
Mercer, Alex, Ronnholm, Harriet, Holmberg, Johan, Lundh, Hanna, Heidrich, Jessica, Zachrisson, Olof, Ossoinak, Amina, Frisen, Jonas, and Patrone, Cesare (2004). 'PACAP promotes neural stem cell proliferation in adult mouse brain'. *Journal of Neuroscience Research* 76: 205-215.
Michel, George F., and Harkins, Debra A. (1985). 'Concordance of handedness between teacher and student facilitates learning manual skills'. *Journal of Human Evolution* 14(6): 597-601.
Miller, Geoffrey (2000). *The Mating Mind. How Sexual Choice Shaped the Evolution of Human Nature*. New York: Doubleday.
Miller, Julie E., Spiteri, Elizabeth, Condro, Michael C., Dosumu-Johnson, Ryan T., Geschwind, Daniel H., and White, Stephanie A. (2008). 'Birdsong decreases protein levels of FoxP2, a molecule required for human speech'. *Journal of Neurophysiology* 1000(4): 2015-2025.
Millikan, Ruth G. (1984). *Language, Thought, and Other Biological Categories*. Cambridge, MA: MIT Press.
—— (1993). *White Queen Psychology and Other Essays for Alice*. Cambridge, MA: MIT Press.
Minelli, Alessandro (1996). 'Some thoughts on homology 150 years after Owen's definition'. *Memorie della Società Italiana di Scienze Naturali e del Museo Civico di Storia Naturale di Milano* XXVII(I) [Systematic Biology as an Historical Science]: 71-79.
—— (1998). 'Molecules, developmental modules, and phenotypes: A combinatorial approach to homology'. *Molecular Phylogenetics and Evolution* 9(3): 340-347.
—— (2000). 'Limbs and tail as evolutionarily diverging duplicates of the main body axis'. *Evolution and Development* 2(3): 157-165.

Minelli, Alessandro (2003). *The Development of Animal Form: Ontogeny, Morphology, and Evolution*. Cambridge: Cambridge University Press.
—— (2007). *Forme del Divenire*. Torino: Einaudi (English version: *Forms of Becoming*. Princeton, NJ: Princeton University Press, 2009).
—— (2009). *Perspectives in Animal Phylogeny and Evolution*. Oxford: Oxford University Press.
—— (2010). 'Evolutionary developmental biology does not offer a significant challenge to the Neo-Darwinian paradigm'. In F. J. Ayala and R. Arp (eds), *Contemporary Debates in Philosophy of Biology*. Chichester: Wiley-Blackwell, 213–226.
—— (2011). 'A principle of developmental inertia'. In B. Hallgrímson and B. K. Hall (eds), *Epigenetics. Linking Genotype and Phenotype in Development and Evolution*. Berkeley, CA: University of California Press, 116–133.
—— and Schram, Frederick R. (1994). 'Owen revisited: A reappraisal of morphology in evolutionary biology'. *Bijdragen tot de Dierkunde* 64: 65–74.
Mitani, John C., and Marler, Peter (1989). 'A phonological analysis of male gibbon singing behavior'. *Behavior* 109: 20–45.
Mithen, Steve (1996). *The Prehistory of Man. A Search for the Origins of Art, Religion, and Science*. London: Thames and Hudson.
Møller, Anders P. (1994). *Sexual Selection and the Barn Swallow*. Oxford: Oxford University Press.
Möller, Ralf (2000). 'Insect visual homing strategies in a robot with analog processing'. *Biological Cybernetics* 83: 231–243
Montaigne, Michel de (1595). 'Apología de Ramón Sibiuda'. In *Los ensayos (según la edición de 1595 de Marie de Gournay)*. Barcelona: Acantilado, translated by J. Bayod Brau, 2007, 628–913 ['An apology for Raymond Sebond'. In *The Complete Essays*. Ed. M. A. Screech. London: Penguin, 1987].
Moore, Allen J. (1988). 'Female preferences, male social status, and sexual selection in *Nauphoeta cinerea*'. *Animal Behavior* 36: 303–305.
Moore, David S. (2001). *The Dependent Gene. The Fallacy of 'Nature vs. Nurture'*. New York: Henry Holt.
Morgan, Thomas H. (1932). *The Scientific Basis of Evolution*. New York: W.W. Norton & Company.
—— Sturtevant, Alfred H., Muller, Hermann J., and Bridges, Calvin B. (1915). *The Mechanism of Mendelian Heredity*. New York: Henry Holt.
Moro, Andrea, Tettamanti, Marco, Perani, Daniela, Donati, Caterina, Cappa, Stefano F., and Fazio, Ferruccio (2001). 'Syntax and the brain: Disentangling grammar by selective anomalies'. *NeuroImage* 13: 110–118.
Müller, Gerd B. (2008). 'EvoDevo as a discipline'. In A. Minelli and G. Fusco (eds), *Evolving Pathways: Key Themes in Evolutionary Developmental Biology*. Cambridge: Cambridge University Press.
Musso, Mariacristina, Moro, Andrea, Glauche, Volkmar, Rijntjes, Michel, Reichenbach, Jürgen, Büchel, Christian, and Weiller, Cornelius (2003). 'Broca's area and the language instinct'. *Nature Neuroscience* 6: 774–781.

Nagel, Thomas (1974). 'What is it like to be a bat?'. *The Philosophical Review* LXXXIII(4): 435–450.
Negri, Toni (2001). 'La linea del mostro'. In U. Fadini, A. Negri, and C. T. Wolfe, (eds), *Desiderio del mostro. Dal circo al laboratorio alla politica*. Rome: Manifestolibri, 7–11.
Nevins, Andrew, Pesetsky, David, and Rodrigues, Cilene (2009a). 'Evidence and argumentation: A reply to Everett (2009)'. *Language* 85(3): 671–681.
—————(2009b). 'Pirahã exceptionality: A reassessment'. *Language* 85(2): 355–404.
Newman, Stuart A. (2003). 'Hierarchy'. In B. K. Hall and W. M. Olson (eds), *Keywords and Concepts in Evolutionary Developmental Biology*. Cambridge, MA: Harvard University Press, 169–174.
Nietzsche, Friedrich (1887). *La ciencia gaya*. Barcelona: Edaf, translated by José Carlos Mardomingo, 2002 [*The Gay Science*. Ed. Bernard Williams. Cambridge: Cambridge University Press, 2001].
Nisbet, Ian C. T. (1973). 'Courtship feeding, egg size and breeding success in common terns'. *Nature* 241: 141–142.
Niyogi, Partha, and Berwick, Robert C. (1995). 'The logical problem of language change'. AI Memo No. 1516, Artificial Intelligence Lab, MIT.
————(1997a). 'The logical problem of language change: A case study of European Portuguese'. *Syntax. A Journal of Theoretical, Experimental and Interdisciplinary Research* 1(2): 195–205.
————(1997b). 'A dynamical systems model for language change'. *Complex Systems* 11: 161–204.
————(1997c). 'Evolutionary consequences of language learning'. *Linguistics & Philosophy* 20(6): 697–719.
————(2009). 'The proper treatment of language acquisition and change in a population setting'. *Proceedings of the National Academy of Sciences USA* 106: 10124–10129.
O'Donald, Peter (1963). 'Sexual selection and territorial behaviour'. *Heredity* 18: 361–364.
O'Donnell, Timothy J., Hauser, Marc D., and Fitch, W. Tecumseh. (2005). 'Using mathematical models of language experimentally'. *Trends in Cognitive Sciences* 9(6): 284–289.
Odell, Garret M., Oster, George F., Alberch, Pere, and Burnside, Beth (1981). 'The mechanical basis of morphogenesis: I. Epithelial folding and invagination'. *Developmental Biology* 85: 446–462.
Odling-Smee, F. John, Laland, Kevin N., and Feldman, Marcus W. (2003). *Niche Construction: The Neglected Process in Evolution*. Princeton, NJ: Princeton University Press.
Okanoya, Kazuo (2002). 'Sexual display as a syntactic vehicle: The evolution of syntax in birdsong and human language through sexual selection'. In A. Wray (ed.), *The Transition to Language*. Oxford: Oxford University Press, 46–64.
Oster, George F., and Alberch, Pere (1982). 'Evolution and bifurcation of developmental programs'. *Evolution* 36, 444–459.
Ouattara, Karim, Lemasson, Alban, and Zuberbühler, Klaus (2009). 'Campbell's monkeys concatenate vocalizations into context-specific call sequences'. *Proceedings of the National Academy of Sciences USA* 106(51), 22026–22031.

Owen, Richard (1843). *Lectures on the Comparative Anatomy and Physiology of the Invertebrate Animals*. London: Longman, Brown, Green, and Longmans.
—— (1848). *On the Archetype and Homologies of the Vertebrate Skeleton*. London: John Van Voorst.
—— (1849). *On the Nature of Limbs. A Discourse*. London: John Van Voorst.
—— (1866). *On the Anatomy of Vertebrates. Vol. I: Fishes and Reptiles*. London: Longmans, Green, and Co.
—— (1868). *On the Anatomy of Vertebrates. Vol. III: Mammals*. London: Longmans, Green, and Co.
Oyama, Susan (2000a). *The Ontogeny of Information*. Second edition. Durham, NC: Duke University Press.
—— (2000b). 'Causal democracy and causal contributions in Developmental Systems Theory'. *Philosophy of Science* 67: S332-S347.
—— Griffiths, Paul E., and Gray, Russell D. (eds) (2001). *Cycles of Contingency. Developmental Systems and Evolution*. Cambridge, MA: MIT Press.
Papadimitriou, Christos H. (1994). *Computational Complexity*. Reading, MA: Addison-Wesley.
Parker, Sue Taylor (2000). 'Comparative developmental evolutionary biology, anthropology, and psychology'. In S. T. Parker, J. Langer, and M. L. McKinney (eds), *Biology, Brains, and Behavior: The Evolution of Human Development*. Santa Fe, NM: School of American Research Press, 1–24.
—— and McKinney, Michael L. (1999). *Origins of Intelligence. The Evolution of Cognitive Development in Monkeys, Apes, and Humans*. Baltimore, MD: Johns Hopkins University Press.
Paterson, Hugh E. H. (1985). 'The recognition concept of species'. In E. S. Vrba (ed.), *Species and Speciation*. Pretoria: Transvaal Museum Monograph No.4: 21–29.
Penrose, Roger (1989). *The Emperor's New Mind: Concerning Computers, Minds, and the Laws of Physics*. Oxford: Oxford University Press.
—— (1994). *Shadows of the Mind: A Search for the Missing Science of Consciousness*. Oxford: Oxford University Press.
Perruchet, Pierre, and Rey, Arnaud (2005). 'Does the mastery of center-embedded linguistic structures distinguish humans from non-humans?'. *Psychonomic Bulletin and Review* 12: 307–313.
Peschke, Klaus, Friedrich, Peter, Gantert, Cornelia, and Metzler, Manfred (1996). 'The use of the tergal gland defensive secretion in male intrasexual aggression of the rove beetle, *Aleochara curtula* (Coleoptera: Staphylinidae), measured by closed-loop-stripping-analyses and tandem bioassay–mass fragmentography'. *Chemoechology* 7: 24–33.
Petersson, Karl Magnus, Folia, Vasiliki, and Hagoort, Peter (2010). 'What artificial grammar learning reveals about the neurobiology of syntax'. *Brain and Language*, doi 10.1016/j.bandl.2010.08.003.
Petrie, Marion, Halliday, Tim, and Sanders, Carolyn (1991). 'Peahens prefer peacocks with elaborate trains'. *Animal Behavior* 41: 323–331.
Pettigrew, John D. (1999). 'Electroreception in monotremes'. *The Journal of Experimental Biology* 202: 1447–1454.

Pfeiffer, John E. (1982). *The Creative Explosion. An Inquiry into the Origins of Art and Religion*. New York: Harper and Row.
Piattelli-Palmarini, Massimo, and Uriagereka, Juan (2005). 'The evolution of the narrow faculty of language'. *Lingue e Linguaggio* IV/1: 27–79.
—— —— (2011). 'A geneticist's dream, a linguist's nightmare: The case of FOXP2'. In A.M. Di Sciullo and C. Boeckx (eds), *The Biolinguistic Enterprise: New Perspectives on the Evolution and Nature of the Human Language Faculty*. Oxford: Oxford University Press, 100–125.
Pigliucci, Massimo (2007). 'Do we need an extended evolutionary synthesis?'. *Evolution* 61(12): 2743–2749.
—— (2010a). 'A misguided attack on evolution'. *Nature* 646: 353–354.
—— (2010b). 'Genotype–Phenotype mapping and the end of the "genes as blueprint" metaphor'. *Philosophical Transactions of the Royal Society of London B* 365: 557–566.
—— and Müller, Gerd B. (2010). *Evolution. The Extended Synthesis*. Cambridge, MA: MIT Press.
Pinel, John P. J., Gorzalka, Boris B., and Ladak, Ferial (2003). 'Cadaverine and putrescine initiate the burial of dead conspecifics by rats'. *Physiology and Behavior* 27(5): 819–824.
Pinker, Steven (1997). *How the Mind Works*. New York: Norton.
—— and Bloom, Paul (1990). 'Natural language and natural selection'. *Behavioral and Brain Sciences* 13: 707–727.
—— and Jackendoff, Ray (2005). 'The faculty of language: What's special about it'. *Cognition* 95: 201–236.
Place, Ullin T. (1956). 'Is consciousness a brain process?'. *British Journal of Psychology* 47: 44–50.
Pliny the Elder (1998). *Historia natural*. Madrid: UNAM/Visor Libros, translated by Francisco Hernández [*Natural History (A selection)*. Ed. John Healey. London: Penguin, 1991].
Ploog, Dieter (2002). 'Is the neural basis of vocalization different in non-human primates and *Homo sapiens*?'. In T. J. Crow (ed.), *The Speciation of Modern Homo sapiens*. London: British Academy, 121–135.
Plotkin, Henry (1997). *Evolution in Mind: An Introduction to Evolutionary Psychology*. London: Alan Lane.
Plutarch (2005). *Vidas de los diez oradores. Sobre la astucia de los animales. Sobre los ríos*. Madrid: Akal, translated by Inmaculada Rodríguez Moreno ['Whether land or sea animals are cleverer'. In *Moralia*, vol. XII. Ed. Harold Cherniss and W. C. Helmbold. Cambridge, MA: Harvard University Press (Loeb Classical Library 406), 1957].
Poeppel, David, and Embick, David (2005). 'Defining the relation between linguistics and neuroscience'. In A. Cutler (ed.), *Twenty-first Century Psycholinguistics: Four Cornerstones*. Mahwah, NJ: Lawrence Erlbaum Associates Inc., 103–118.
Ponce de León, Marcia S., Golovanova, Lubov, Doronichev, Vladimir, Romanova, Galina, Akazawa, Takeru, Kondo, Osamu, Ishida, Hajime, Zollikofer, Christoph P. E. (2008). 'Neanderthal brain size at birth provides insights into the evolution of human life history'. *Proceedings of the National Academy of Sciences USA* 105(37): 13764–13768.
Port, Robert F., and van Gelder, Timothy (eds) (1995). *Mind as Motion. Explorations in the Dynamics of Cognition*. Cambridge, MA: MIT Press.

Premack, David (1985). '"Gavagai!" or the future history of the animal language controversy'. *Cognition* 19: 207-296.
Proctor, Heather C. (1996). 'Behavioral characters and homoplasy: Perception versus practice'. In M. J. Sanderson and L. Hufford (eds), *Homoplasy: The Recurrence of Similarity in Evolution*. San Diego: Academic Press, 131-149.
Ptak, Susan, Enard, Wolfgang, Wiebe, Victor, Hellmann, Ines, Krause, Johannes, Lachmann, Michael, and Pääbo, Svante (2009). 'Linkage desequilibrium extends across putative selected sites in FOXP2'. *Molecular Biology and Evolution* 26(10): 2181-2184.
Pullum, Geoffrey K. (1986). 'On the relations of IDC-command and government'. In *Proceedings of the West Coast Conference on Formal Linguistics. Vol. 5*. Stanford, CA: The Stanford Linguistics Association, 92-206.
Pulvermüller, Friedemann, and Assadollahi, Ramin (2007). 'Grammar or serial order? Discrete combinatorial brain mechanisms reflected by the Syntactic Mismatch Negativity'. *Journal of Cognitive Neuroscience* 19: 971-980.
Pusey, Anne E., and Packer, Craig (1994). 'Infanticide in lions: Consequences and counterstrategies'. In S. Parmigiani and F. von Saal (eds), *Infanticide and Parental Care*. London: Harwood, 279-299.
Putnam, Hilary (1975 [1960]). 'Minds and machines'. In *Mind, Language, and Reality. Philosophical Papers, Vol. 2*. Cambridge: Cambridge University Press, 362-385.
—— (1988). *Representation and Reality*. Cambridge, MA: MIT Press.
Pylyshyn, Zenon W. (1980). 'Computation and cognition: Issues in the foundation of Cognitive Science'. *Brain and Behavioral Sciences* 3: 111-132.
—— (1984). *Computation and Cognition: Toward a Foundation of Cognitive Science*. Cambridge, MA: MIT Press.
Radzinski, Daniel (1991). 'Chinese number-names, tree adjoining languages, and mild context-sensitivity'. *Computational Linguistics* 17: 277-299.
Raff, Rudolf A. (2000). 'Evo-devo: The evolution of a new discipline'. *Nature Reviews Genetics* 1: 74-79.
Rakic, Pasko, and Kornack, David R. (2001). 'Neocortical expansion and elaboration during primate evolution'. In D. Falk and K. R. Gibson (eds), *Evolutionary Anatomy of the Primate Cerebral Cortex*. Cambridge: Cambridge University Press, 30-56.
Rampino, Michael R., and Ambrose, Stanley H. (2000). 'Volcanic winter in the Garden of Eden: The Toba super-eruption and Late Pleistocene human population bottleneck'. *Geological Society of America Special Papers* 345: 71-82.
Rasskin-Gutman, Diego (2005). 'Modularity: Jumping forms within morphospace'. In W. Callebaut and D. Rasskin-Gutman (eds), *Modularity: Understanding the Development and Evolution of Complex Systems*. Cambridge, MA: MIT Press, 207-219.
—— (2009). 'Molecular evo-devo: The path not taken by Pere Alberch'. In D. Rasskin-Gutman and M. De Renzi (eds), *Pere Alberch. The Creative Trajectory of an Evo-Devo Biologist*. Valencia: Institut d'Estudis Catalans and Universitat de València, 67-84.
—— and De Renzi, Miquel (eds) (2009). *Pere Alberch. The Creative Trajectory of an Evo-Devo Biologist*. Valencia: Institut d'Estudis Catalans and Universitat de València.
Reid, Robert G. B. (2007). *Biological Emergences. Evolution by Natural Experiment*. Cambridge, MA: MIT Press.

Reid, Thomas (1764). *An Inquiry into the Human Mind on the Principles of Common Sense*. Ed. Derek R. Brookes. Edinburgh: Edinburgh University Press, 1997.

Reiner, Anton (2010). 'The conservative evolution of the vertebrate basal ganglia'. In H. Steinzer and K. Tseng (eds), *Handbook of Basal ganglia Structure and Function*. London: Academic Press, 29–62.

—— Brauth, Steven E., and Karten, Harvey J. (1984). 'Evolution of the amniote basal ganglia'. *Trends in Neurosciences* 7: 320–325.

Reiss, John O., Burke, Ann C., Archer, Charles, De Renzi, Miquel, Dopazo, Hernán, Etxeberria, Arantza, Gale, Emily A., Hinchliffe, Richard, Nuño de la Rosa García, Laura, Rose, Chris S., Rasskin-Gutman, Diego, and Müller, Gerd B. (2009). 'Pere Alberch: Originator of EvoDevo'. *Biological Theory* 3(4): 351–356.

Rendall, Drew, and Di Fiore, Anthony (2007). 'Homoplasy, homology, and the perceived special status of behavior in evolution'. *Journal of Human Evolution* 52: 504–521.

Rice, Sean H. (2008). 'Theoretical approaches to the evolution of development and genetic architecture'. *Annals of the New York Academy of Sciences* 1133: 67–86.

Richards, Evelleen (1994). 'A political anatomy of monsters, hopeful and otherwise: Teratogeny, transcendentalism, and evolutionary thinking'. *Isis* 85(3): 377–411.

Richards, Marc (2008). 'Two kinds of variation in a minimalist system'. *Linguistische Arbeits Berichte* 87: 133–162.

Richards, Robert J. (1987). *Darwin and the Emergence of Evolutionary Theories of Mind*. Chicago: University of Chicago Press.

Ridley, Matt (1993). *The Red Queen. Sex and the Evolution of Human Nature*. New York: Perennial.

Rieppel, Olivier (1992). 'Homology and logical fallacy'. *Journal of Evolutionary Biology* 5: 701–715.

—— (1994). 'Homology, topology, and typology: The history of modern debates'. In B. K. Hall (ed.), *Homology: The Hierarchical Basis of Comparative Biology*. San Diego: Academic Press, 63–100.

Ristad, Eric Sven (1993). *The Language Complexity Game*. Cambridge, MA: MIT Press.

Robert, Jason S. (2004). *Embryology, Epigenesis, and Evolution. Taking Development Seriously*. Cambridge: Cambridge University Press.

—— Hall, Brian K., and Olson, Wendy M. (2001). 'Bridging the gap between developmental systems and evolutionary developmental biology'. *BioEssays* 23: 954–962.

Rochefort, Christele, He, Xialou, Scotto-Lomassese, Sophie, and Scharff, Constance (2007). 'Recruitment of FoxP2-expressing neurons to Area X varies during song development'. *Developmental Neurobiology* 67(6): 809–817.

Roelink, Henk, Augsburger, A., Heemskerk, J., Korzh, Vladimir, Norlin, S., Ruiz i Altaba, Ariel, Tanabe, Y., Placzek, Marysia, Edlund, T., Jessell, Thomas M., and Dodd, J. (1994). 'Floor plate and motor neuron induction by vhh-1, a vertebrate homolog of hedgehog expressed by the notochord'. *Cell* 76: 761–775.

Roessler, Erich, Belloni, Elena, Gaudenz, Karin, Jay, Philippe, Berta, Philippe, Scherer, Stephen W., Tsui, Lap-Chee, and Muenke, Maximilian (1996). 'Mutations in the human *Sonic Hedgehog* gene cause holoprosencephaly'. *Nature Genetics* 14: 357–360.

Rogers, James, and Hauser, Marc D. (2010). 'The use of Formal Language Theory in studies of artificial language learning: A proposal for distinguishing the differences between human and nonhuman animal learners'. In Harry van der Hulst (ed.), *Recursion and Human Language*. New York: de Gruyter, 213–232.

—— and Pullum, Geoffrey K. (2011). 'Aural pattern recognition experiments and the Subregular Hierarchy'. *Journal of Logic, Language and Information* 20: 329–342.

Roth, V. Louise (1984). 'On homology'. *Biological Journal of the Linnean Society* 22: 13–29.

—— (1988). 'The biological basis of homology'. In C. J. Humphries (ed.), *Ontogeny and Systematics*. New York: Columbia University Press, 1–26.

—— (1991). 'Homology and hierarchies: Problems solved and unresolved'. *Journal of Evolutionary Biology* 4: 167–194.

Rousseau, Jean-Jacques (1755). *Discours sur l'origine et les fondements de l'inégalité parmi les hommes* (edited by J. Starobinski). Paris: Gallimard, 1985 ['Discourse on the origin and foundations of inequality among men'. In *The* Discourses *and other Early Political Writings*. Ed. Victor Gourevitch. Cambridge: Cambridge University Press, 1997].

—— (1781). *Essai sur l'origine des langues* (edited by J. Starobinski). Paris: Gallimard, 1990 ['Essay on the origin of language'. In *The* Discourses *and other Early Political Writings*. Ed. Victor Gourevitch. Cambridge: Cambridge University Press, 1997].

Rubel, Lee A. (1985). 'The brain as an analog computer'. *Journal of Computational Neurobiology* 4: 73–81.

Rupke, Nicolaas A. (1994). *Richard Owen: Victorian Naturalist*. New Haven, CT: Yale University Press.

Ruse, Michael (2003). *Darwin and Design. Does Evolution Have a Purpose?* Cambridge, MA: Harvard University Press.

Russell, Edward S. (1916). *Form and Function: A Contribution to the History of Animal Morphology*. London: Murray.

Ryle, Gilbert (1949). *The Concept of Mind*. London: Hutchinson.

Salie, Rishard, Niederkofler, Vera, and Arber, Silvia (2005). 'Patterning molecules, multitasking in the nervous system'. *Neuron* 45: 189–192.

Sanderson, M. J., Baldwin, B. G., Bharathan, G., Campbell, C. S., Von Dohlen, C., Ferguson, D., Porter, J. M., Wojciechowski,, M. F., and Donoghue, M. J. (1993). 'The growth of phylogenetic information and the need for a phylogenetic data base'. *Systematic Biology* 42: 562–568.

Sarpeshkar, Rahul (1998). 'Analog versus digital: Extrapolating from electronics to neurobiology'. *Neural Computation* 10: 1601–1638.

—— (2009). 'Neuromorphic and biomorphic engineering systems'. In *McGraw-Hill Yearbook of Science & Technology 2009*. New York: McGraw-Hill, 250–252.

Saussure, Ferdinand de (1916). *Cours de linguistique générale*. Ed. Charles Bally and Albert Sechehaye. Paris: Payot [*Course in General Linguistics*. Ed. Roy Harris. La Salle, IL: Open Court, 1983].

Savage-Rumbaugh, E. Sue, Murphy, Jeannine, Sevcik, Rose A., Brakke, Karen E., Williams, Shelly L., Rumbaugh, Duane M., and Bates, Elizabeth (1993). 'Language comprehension in ape and child'. *Monographs of the Society for Research in Child Development*, serial no. 233, vol. 58, nos. 3/4.

—— Shanker, Stuart G., and Taylor, Talbot J. (1998). *Apes, Language, and the Human Mind*. Oxford: Oxford University Press.

Scheutz, Matthias (1999). 'When physical systems realize functions...'. *Minds and Machines* 9: 161–196.

—— (2002). 'Computationalism—The next generation'. In M. Scheutz (ed.), *Computationalism. New Directions*. Cambridge, MA: MIT Press, 1–21.

Schierwater, Bernd, and DeSalle, Rob (2001). 'Current problems with the zootype and the early evolution of Hox genes'. *Journal of Experimental Zoology* 291: 169–174.

Schmidt-Rhaesa, Andreas (2007). *The Evolution of Organ Systems*. Oxford: Oxford University Press.

Schoenauer, Norbert (1981). *6000 Years of Housing*. New York: Garland.

Searle, John R. (1980). 'Minds, brains, and programs'. *Behavioral and Brain Sciences* 3: 417–424.

—— (1992). *The Rediscovery of the Mind*. Cambridge, MA: MIT Press.

—— (1995). *The Construction of Social Reality*. New York: The Free Press.

Shannon, Claude, and Weaver, Warren (1949). *The Mathematical Theory of Communication*. Urbana, IL: The University of Illinois Press.

Shtyrov, Yuri, Pulvermüller, Friedemann, Näätänen, Risto, and Ilmoniemi, Risto J. (2003). 'Grammar processing outside the focus of attention: An MEG study'. *Journal of Cognitive Neuroscience* 15: 1195–1206.

Shu, Yousheng, Hasenstaub, Andrea, Duque, Álvaro, Yu, Yuguo, and McCormick, David (2006). 'Modulation of intracortical synaptic potentials by presynaptic somatic membrane potential'. *Nature* 441: 761–765.

Shubin, Neil (2008). *Your Inner Fish. The Amazing Discovery of Our 375-Million-Year-Old Ancestor*. New York: Pantheon Books.

—— Tabin, Cliff, and Carroll, Sean (1997). 'Fossils, genes and the evolution of animal limbs'. *Nature* 388: 639–648.

—— —— —— (2009). 'Deep homology and the origins of evolutionary novelty'. *Nature* 457: 818–823.

Simon, Herbert (1969). *The Sciences of the Artificial*. Cambridge, MA: MIT Press.

Simpson, George G. (1944). *Tempo and Mode in Evolution*. New York: Columbia University Press.

—— (1961). *Principles of Animal Taxonomy*. New York: Columbia University Press.

Singer, Tatjana, McConnell, Michael J., Marchetto, Maria C. N., Coufal, Nicole G., and Gage, Fred H. (2010). 'LINE-1 retrotransposons: Mediators of somatic variation in neuronal genomes?'. *Trends in Neurosciences* 33: 345–354.

Skinner, Burrhus F. (1938). *The Behavior of Organisms*. New York: Appleton-Century-Crofts.

—— (1953). *Science and Human Behavior*. New York: Macmillan.

—— (1957). *Verbal Behavior*. New York: Appleton-Century-Crofts.

—— (1975). 'The shaping of phylogenic behavior'. *Journal of the Experimental Analysis of Behavior* 24: 117–120.

Slack, Jonathan M. W., Holland, Peter W. H., and Graham, Christopher F. (1993). 'The zootype and the phylotypic stage'. *Nature* 361: 490–492.

Smart, John J. C. (1959). 'Sensations and brain processes'. *Philosophical Review* 68: 141–156.

Smith, Brian Cantwell (2002). 'The foundations of computing'. In M. Scheutz (ed.), *Computationalism. New Directions*. Cambridge, MA: MIT Press, 23–58.

—— (2011). 'Introduction'. In *Age of Significance*. Vol. I. Available online at <http://www.ageofsignificance.org>.

Smith, John W. (1977). *The Behavior of Communicating: An Ethological Approach*. Cambridge, MA: Harvard University Press.

Smith, Kathleen K. (2001). 'Heterochrony revisited: The evolution of developmental sequences'. *Biological Journal of the Linnean Society* 73: 169–186.

—— (2002). 'Sequence heterochrony and the evolution of development'. *Journal of Morphology* 252: 82–97.

Smith, Kenny, and Kirby, Simon (2008). 'Cultural evolution: Implications for understanding the human language faculty and its evolution'. *Philosophical Transactions of the Royal Society B* 363: 3591–3603.

Smolensky, Paul (1988). 'On the proper treatment of connectionism'. *Behavioral and Brain Sciences* 11: 1–23.

—— (1995). 'Connectionism, constituency and the language of thought'. In C. Macdonald and G. Macdonald (eds), *Connectionism*. Oxford: Blackwell, 164–198.

Sober, Elliott (1980). 'Evolution, population thinking, and essentialism'. *Philosophy of Science* 47: 350–383.

Soffer, Olga, Adovasio, James M., and Hyland D. C. (2000). 'The "Venus" figurines. Textiles, basketry, gender, and status in the Upper Paleolithic'. *Current Anthropology* 41: 511–525.

Spiteri, Elisabeth, Konopka, Genevieve, Coppola, Giovanni, Bomar, Jamee, Oldham, Michael, Ou, Jing, Vernes, Sonja C., Fisher, Simon E., Ren, Bing, and Geschwind, Daniel H. (2007). 'Identification of the transcriptional targets of *FOXP2*, a gene linked to speech and language, in developing human brain'. *American Journal of Human Genetics* 81: 1144–1157.

Stevens, Peter F. (1984). 'Homology and phylogeny: Morphology and systematics'. *Systematic Botany* 9: 395–409.

Stich, Stephen P. (1975). 'The idea of innateness'. In S. P. Stich (ed.), *Innate Ideas*. Berkeley, CA and London: University of California Press, 1–22 (Reprinted in S. P. Stich (2011). *Collected Papers. Volume 1: Mind and Language, 1972–2010*. Oxford: Oxford University Press, 20–35).

Strausfeld, H. J., Homburg, U., and Kloppenberg, P. (2000). 'Parallel organization in honey bee mushroom bodies by peptidergic Kenyon cells'. *Journal of Comparative Neurology* 424: 179–195.

Striedter, Georg F. (2005). *Principles of Brain Evolution*. Sunderland, MA: Sinauer.

—— Marchant, T. Alejandro, and Beydler, Sarah (1998). 'The "neostriatum" develops as part of the lateral pallium in birds'. *Journal of Neurosciences* 18: 5839–5849.

—— and Northcutt, R. Glenn (1991). 'Biological hierarchies and the concept of homology'. *Brain, Behavior and Evolution* 38: 177–189.

Struhsaker, Thomas T. (1967). 'Auditory communication among vervet monkeys (*Cercopithecus aethiops*)'. In S. A. Altmann (ed.), *Social Communication among Primates*. Chicago: University of Chicago Press: 1197–1203.

Surbey, Michele K. (2008). 'Selfish genes, developmental systems, and the evolution of development'. In Ch. Crawford and D. Krebs (eds), *Foundations of Evolutionary Psychology*. New York: Lawrence Erlbaum, 137–149.

Takahashi, Kaori, Liu, Fu-Chin, Hirokawa, Katsuiku, and Takahashi, Hiroshi (2003). 'Expression of *Foxp2*, a gene involved in speech and language, in the developing and adult striatum'. *Journal of Neuroscience Research* 73: 61–72.

Tattersall, Glenn J., Andrade, Denis V., and Abe, Augusto S. (2009). 'Heat exchange from the toucan bill reveals a controllable vascular thermal radiator'. *Science* 325: 468–470.

Teramitsu, Ikuko, Kudo, Lili C., London, Sarah, Geschwind, Daniel H., and White, Stephanie A. (2004). 'Parallel *FoxP1* and *FoxP2* expression in songbird and human brain predicts functional interaction'. *Journal of Neuroscience* 24(23): 3152–3163.

—— and White, Stephanie A. (2006). '*FoxP2* regulation during undirected singing in adult songbirds'. *Journal of Neuroscience* 26: 7390–7394.

Tesnière, Lucien V. (1976). *Éléments de syntaxe structurale*. Paris: Éditions Klincksieck.

Theißen, Günter (2009). 'Saltational evolution: Hopeful monsters are here to stay'. *Theory in Biosciences* 128: 43–51.

Thelen, Esther, and Smith, Linda B. (1994). *A Dynamic Systems Approach to the Development of Cognition and Action*. Cambridge, MA: MIT Press.

Theraulaz, Guy, and Bonabeau, Eric (1995). 'Modelling the collective building of complex architectures in social insects with lattice swarms'. *Journal of Theoretical Biology* 177: 381–400.

—— —— and Deneubourg, Jean-Louis (1998). 'The origin of nest complexity in social insects'. *Complexity* 3(6): 15–25.

Thornhill, Randy (1992). 'Female preference for the pheromone of males with low fluctuating asymmetry in the Japanese scorpionfly (*Panorpa japonica*, Mecoptera)'. *Behavioral Ecology* 3: 277–283.

Tinbergen, Niko (1951). *A Study of Instinct*. Oxford: Clarendon Press.

Tooby, John, and DeVore, Irven (1987). 'The reconstruction of hominid behavioral evolution through strategic modeling'. In W. G. Kinzey (ed.), *The Evolution of Human Behavior: Primate Models*. Albany, NY: SUNY Press, 183–237.

Townsend, David J., and Bever, Thomas G. (2001). *Sentence Comprehension: The Integration of Habits and Rules*. Cambridge, MA: MIT Press.

Tracy, Joseph, Flanders, Adam, Madi, Saussan, Laskas, Joseph, Stoddard, Eve, Pyrros, Ayis, Natale, Peter, and DelVecchio, Nicole (2003). 'Regional brain activation associated with different performance patterns during learning of a complex motor skill'. *Cerebral Cortex* 13(9): 904–910.

Ujhelyi, Mária (1996). 'Is there any intermediate stage between animal communication and language?'. *Journal of Theoretical Biology* 180: 71–76.

—— (1998). 'Long-call structure in apes as a possible precursor for language'. In J. R. Hurford, M. Studdert-Kennedy, and C. Knight (eds), *Approaches to the Evolution of Language. Social and Cognitive Bases*. Cambridge: Cambridge University Press, 177–189.

Ullman, Michael T. (2004). 'Contributions of memory circuits to language: The declarative/procedural model'. *Cognition* 92: 231–270.

Uriagereka, Juan (2009). 'Uninterpretable features in syntactic evolution'. In M. Piattelli-Palmarini, J. Uriagereka, and P. Salaburu (eds), *Of Minds and Language. A Dialogue with Noam Chomsky in the Basque Country*. Oxford: Oxford University Press, 169–183.

Valentine, James W. (2004). *On the Origin of Phyla*. Chicago: The University of Chicago Press.

van Gelder, Timothy (1995). 'What might cognition be, if not computation?'. *The Journal of Philosophy* 92: 345–381.
—— (1998). 'The dynamical hypothesis in cognitive science'. *Behavioral and Brain Sciences* 21: 615–661.
van Heijningen, Caroline A. A., de Visser, Jos, Zuidema, Willem, and ten Cate, Carel (2009). 'Simple rules can explain discrimination of putative recursive syntactic structures by a songbird species'. *Proceedings of the National Academy of Sciences USA* 106(48): 20538–20543.
Van Valen, Leigh M. (1982). 'Homology and causes'. *Journal of Morphology* 173: 305–312.
Vanhaeren, Marian, d'Errico, Francesco, Stringer, Chris, James, Sarah L., Todd, Jonathan A., and Mienis, Henk K. (2006). 'Middle paleolithic shell beads in Israel and Algeria'. *Science* 312: 1785–1788.
Vicq d'Azyr, Félix (1774). 'Mémoire sur les rapports qui se trouvent entre les usages et la structure des quatre extrémités dans l'homme et dans les quadrupèdes'. *Histoire de l'Académie Royale des Sciences (1778)*. Paris: Imprimerie Royale, 254–270.
Vijay-Shanker, K., and Weir, David (1994). 'The equivalence of four extensions of context-free grammars'. *Mathematical Systems Theory* 27: 511–546.
Waddington, Conrad H. (1957). *The Strategy of the Genes*. London: Allen and Unwin.
Wagensberg, Jorge (2004). *La rebelión de las formas o cómo perseverar cuando la incertidumbre aprieta*. Barcelona: Tusquets.
Wagner, Andreas (2005). *Robustness and Evolvability in Living Systems*. Princeton, NJ: Princeton University Press.
Wagner, Günter P. (1989a). 'The biological homology concept'. *Annual Review of Ecology and Systematics* 20: 51–69.
—— (1989b). 'The origin of morphological characters and the biological basis of homology'. *Evolution* 43: 1157–1171.
—— (1996). 'Homologues, natural kinds and the evolution of modularity'. *American Zoologist* 36: 36–43.
Wake, David B. (1994). 'Comparative terminology'. *Science* 265: 268–269.
—— (2003). 'Homology and homoplasy'. In Brian K. Hall and Wendy M. Olson (eds), *Keywords and Concepts in Evolutionary Developmental Biology*. Cambridge, MA: Harvard University Press, 191–200.
Wang, Yin-qiu, and Su, Bing (2004). 'Molecular evolution of *microcephalin*, a gene determining human brain size'. *Human Molecular Genetics* 13: 1131–1137.
—— Qian, Ya-ping, Yang, Su, Shi, Hong, Liao, Cheng-hong, Zheng, Hong-Kun, Wang, Ju, Lin, Alice A, Cavalli-Sforza, L. Luca, Underhill, Peter A., Chakraborty, Ranajit, Jin, Li, and Su, Bing (2005). 'Accelerated evolution of the pituitary adenylate cyclase-activating polypeptide precursor gene during human origin'. *Genetics* 170: 801–806.
Webber, Adam Brooks (2008). *Formal Language. A Practical Introduction*. Wilsonville, OR: Franklin, Beedle & Associates.
Webster, Gerry, and Goodwin, Brian (1996). *Form and Transformation. Generative and Relational Principles in Biology*. Cambridge: Cambridge University Press.
Weir, David (1992). 'A geometric hierarchy beyond context-free languages'. *Theoretical Computer Science* 104: 235–261.

—— (1994). 'Linear iterated pushdowns'. *Computational Intelligence* 10: 431–439.
Wenzel, John W. (1992). 'Behavioral homology and phylogeny'. *Annual Review of Ecology and Systematics* 23: 361–381.
—— (1993). 'Application of the biogenetic law to behavioral ontogeny: A test using nest architecture in paper wasps'. *Journal of Evolutionary Biology* 6: 229–247.
West-Eberhard, Mary Jane (2003). *Developmental Plasticity and Evolution*. Oxford: Oxford University Press.
Wiener, Norbert (1948). *Cybernetics or Control and Communication in the Animal and the Machine*. New York: John Wiley & Sons Inc.
Wilson, Edward O. (1975). *Sociobiology. The New Synthesis*. Cambridge, MA: The Belknap Press.
—— (1998). *Consilience. The Unity of Knowledge*. London: Little Brown.
Wilson, Robert A. (1994). 'Wide computationalism'. *Mind* 103(411): 351–372.
—— Barker, Matthew J., and Brigandt, Ingo (2007). 'When traditional essentialism fails: Biological natural kinds'. *Philosophical Topics* 35: 189–215.
Wittgenstein, Ludwig (1922). *Tractatus Logico-Philosophicus*. London: Kegan Paul, Trench and Co.
—— (1953). *Philosophical Investigations/Philosophische Untersuchungen*. Oxford: Basil Blackwell.
Wolpert, David H., and Macready, William G. (1997). 'No free lunch theorems for optimization'. *IEEE Transactions on Evolutionary Computation* 1: 67–82.
Woods, C. Geoffrey (2004). 'Human microcephaly'. *Current Opinion in Neurobiology* 14: 1–6.
Wouters, Arno (1999). *Explanation Without a Cause*. PhD Dissertation, University of Utrecht.
—— (2003). 'Four notions of biological function'. *Studies in History and Philosophy of Biological and Biomedical Sciences* 34: 633–668.
Wright, Larry (1973). 'Functions'. *Philosophical Review* 82: 139–168.
Wynn, Thomas, and Coolidge, Frederick L. (2011). 'The implications of the working memory model for the evolution of modern cognition'. *International Journal of Evolutionary Biology* 2011, Article ID 741357, 12 pages, doi:10-4061/2011/741357.
Wynne-Edwards, Vero C. (1962). *Animal Dispersion in Relation to Social Behaviour*. Edinburgh: Oliver and Boyd.
Young, Bruce A. (1993). 'On the necessity of an archetypal concept in morphology: With special reference to the concepts of "structure" and "homology"'. *Biology and Philosophy* 8: 225–248.
Yu, Fuli, Hill, R. Sean, Schaffner, Stephen F., Sabeti, Pardis C., Wang, Eric T., Mignault, Andre A., Ferland, Russell J., Moyzis, Robert K., Walsh, Christopher A., and Reich, David (2007). 'Comment on "Ongoing adaptive evolution of *ASPM*, a brain size determinant in *Homo sapiens*"'. *Science* 316: 370.
Zahavi, Amotz (1975). 'Mate selection: A selection for a handicap'. *Journal of Theoretical Biology* 53: 205–214.
—— (1977). 'The cost of honesty (further remarks on the handicap principle)'. *Journal of Theoretical Biology* 67: 603–605.
—— and Zahavi, Avishag (1997). *The Handicap Principle. A Missing Piece of Darwin's Puzzle*. Oxford: Oxford University Press.

Zelditch, Miriam L., and Fink, William L. (1996). 'Heterochrony and heterotopy: Stability and innovation in the evolution of form'. *Paleobiology* 22: 241–254.

Zhang, Jianzhi (2003). 'Evolution of the human *ASPM* gene, a major determinant of brain size'. *Genetics* 165: 2063–2070.

Zilhão, João, Angelucci, Diego E., Badal-García, Ernestina, d'Errico, Francesco, Daniel, Floréal, Dayet, Laure, Douka, Katerina, Higham, Thomas F. G., Martínez-Sánchez, María José, Montes-Bernárdez, Ricardo, Murcia-Mascarós, Sonia, Pérez-Sirvent, Carmen, Roldán-García, Clodoaldo, Vanhaeren, Marian, Villaverde, Valentín, Wood, Rachel, and Zapata, Josefina (2010). 'Symbolic use of marine shells and mineral pigments by Iberian Neandertals'. *Proceedings of the National Academy of Sciences USA* 107(3): 1023–1028.

Zylberberg, Ariel, Dehaene, Stanislas, Roelfsema, Pieter, and Sigman, Mariano. (2011). 'The human Turing machine: A neural framework for mental programs'. *Trends in Cognitive Sciences* 15(7): 293–300.

Index of Authors

Aboitiz, Francisco 22, 102, 102 n6
Adams, Colin C. 148 n12
Agol, Ian 148
Aho, Alfred V. 95 n2
Aitchinson, Jane 59
Alba, David B. 115, 127
Alberch, Jordi 117
Alberch, Pere xii, 22, 29–30, 109–11, 110 n1, 111 n4, 116–19, 116 n15, 116 n16, 121–3, 123 n18, 125, 131 n22
Alexander, Richard D. 74 n15
Alle, Henrik 172
Ambrose, Stanley H. 34–35
Amundson, Ronald A. 70 n8, 77 n20, 82, 110, 112
Arendt, Detlev 104
Aristotle 47, 48 n4, 70, 70 n8, 81, 115 n14
Armstrong, David M. 76 n18, 76 n19, 78 n21
Arnold, Kate 50 n7, 141 n3, 143
Arsuaga, Juan Luis 154
Arthur, Wallace 109, 112, 115, 126 n20
Assadollahi, Ramin 102 n6
Atkinson, Quentin D. 35
Atz, James W. 65–6
Avian Brain nomenclature Consortium 106 n13
Avital, Eytan 18

Baddeley, Alan 157 n24
Baker, Mark C. 10
Balari, Sergio xiii, 5, 8, 18, 24, 28, 37, 41, 60–2, 60 n12, 81–2, 89, 104 n10, 130, 147, 149 n13, 154–6, 157 n24, 158
Bar-Hillel, Yehoshua 54 n10
Barham, Lawrence S. 27 n1, 153 n15
Barlow, George W. 76
Barret, H. Clark 132 n26

Barton, Robert A. 131
Bateson, Gregory 32, 79
Bateson, William 29, 125
Bayman, Benjamin F. 149 n13
Behar, Doron M. 24, 35
Benítez-Burraco, Antonio xii, 102 n8, 113 n9, 114, 129, 155
Benveniste, Émile 14 n5
Berwick, Robert C. 3, 19, 93
Bever, Thomas G. 1, 18 n9
Bickerton, Derek 21, 36, 107, 125, 135
Biederman, Irving 150
Blake, Barry J. 15
Blanco, María José 117
Block, Ned 132 n26
Bloom, Paul 19, 111
Bock, Gregory R. 65, 74
Boeckx, Cedric xii, 3, 5, 12
Bonabeau, Eric 169–70
Bonato, Lucio 70 n9
Bond, Jacquelyn 130
Bonner, John Tyler 112
Borgia, Gerald 58, 73, 146 n9
Botha, Rudolf 154 n19
Bourikas, Dimitris 130
Bouzouggar, Abdeljalil 156 n20
Boyden, Alan 66, 83
Braitenberg, Valentino 78 n22
Brigandt, Ingo 83, 85 n27, 86, 86 n28
Brockmann, H. Jane 75
Brooks, Alison S. 27, 27 n1, 153 n15
Brown, William M. 111 n3
Bullock, Theodore H. 133
Burchell, Graham 31 n3
Burish, Mark J. 146 n9
Buss, David M. 111
Byers, John A. 76

Cabanis, Pierre-Jean 88
Camazine, Scott 169–70
Camps, Marta xii, 23, 28, 146–7, 154
Canguilhem, Georges xiv, 29
Cardew, Gail 65, 74
Cardín, Alberto 29
Carroll, Sean B. 22, 109, 113, 113 n6, 113 n7, 114, 116
Casati, Roberto 150
Cave, Carolyn Baker 151
Chalmers, David J. 16 n8, 104 n9
Cheney, Dorothy L. 55, 60, 141
Chierchia, Gennaro 8
Chihara, Charles 77 n19,
Ching, Yick-Pang 130
Chomsky, Noam 1–3, 6, 8, 13, 15–16, 21–22, 42, 53–5, 59, 72 n13, 76 n17, 78 n21, 91–2, 95–7, 95 n2, 104 n10, 107, 125, 135, 139, 174, 176, 184
Christy, John H. 58, 73
Churchland, Paul M. 57, 165
Clark, Andy 15, 16 n8,
Claudius Aelianus 48
Collias, Elsie C. 23, 146, 151–2
Collias, Nicholas E. 23, 146, 151–2
Collins, Sarah 55
Conard, Nicholas J. 156
Coolidge, Frederick L. 156 n23, 157 n24
Coop, Graham 155
Copeland, B. Jack 162 n1
Coqueugniot, Hélène 157
Corballis, Michael C. 3, 136
Corbett, Greville G. 15
Cordemoy, Géraud de 48
Cosmides, Leda 75, 80
Costa, Daniel P. 104
Coughlin, Bridget C. 115 n13
Coyne, Jerry A. 32
Cummings, Jeffrey L. 100
Curry, Haskell B. 12 n4

d'Errico, Francesco 27–28, 28 n2, 153 n15, 153 n16, 154 n19, 156, 156 n21, 156 n23
Darlington, Richard B. 128, 129 n21, 131

Darwin, Charles 19, 29, 46, 48–50, 49 n5, 65–6, 82, 84, 87, 151
Darwin, Erasmus 88
Davies, Nick B. 73–5, 78–9
Davis, Martin 162 n1
Dawkins, Richard 18, 45 n2, 50, 74–5
de Beer, Gavin 82 n25, 87, 115
de Queiroz, Alan 70
De Renzi, Miquel 116–17, 116 n15
de Vries, Hugo 29
Deacon, Terrence 133, 157
Dennett, Daniel C. 57, 60 n12, 77, 77 n20, 79–80, 162, 162 n1, 163 n2, 172
DeSalle, Rob 106
Descartes, René 2, 48
Deutsch, Jean 106 n14, 166 n8
DeVore, Irven 75
Di Fiore, Anthony 70
Dicicco-Bloom, Emanuel 130
Dobzhansky, Theodosius 110–11
Dorus, Steve 130
Dunbar, Robin I. M. 59

Ebbesson, Sven O. E. 133, 157
Ebeling, K. Smilla 58
Echelard, Yann 130
Edgar, Blake 156
Edwards, Laura 114
Eibl-Eibesfeldt, Iräneus 45
Embick, David 4, 102
Emery, Nathan J. 146 n9
Enard, Wolfgang 155
Engels, Friedrich 47
Etxeberria, Arantza 117
Evans, Patrick D. 130
Everett, Daniel 9

Fabre, Jean-Henri Casimir 151
Fadini, Ubaldo 29
Farris, Sarah M. 166–9, 168 n12
Fay, Justin C. 35
Ferland, Russell J. 102 n8
Fink, William L. 126 n20
Finlay, Barbara L. 110, 127–9, 129 n21, 131

Index of Authors

Finlayson, Clive 157
Fitch, W. Tecumseh 3, 92–3, 136, 137 n2, 143 n4, 158
Fodor, Jerry A. 25, 58–9, 72, 76, 77 n19, 78 n21, 79, 91 n1, 110 n2, 114 n11, 132, 132 n26, 139, 161–4, 162 n1, 163 n2, 164 n3, 164–165 n4, 165 n6, 170–171
Foley, R. A. 35
Fortnow, Lance 184
Foucault, Michael 29, 31
Friederici, Angela D. 102 n6
Frith, Clifford B. 146 n9
Futuyma, Douglas J. 32–3

Gallistel, C. Randy 6, 91, 91 n1
Gao, Yijie 130
García-Azkonobieta, Tomás 109
García, Ricardo 102 n6
Garland, Jr, Theodore 104–5
Gathorne-Hardy, F.J. 34
Gazdar, Gerald 12
Geiger, Jörg R. P. 172
Geissmann, Thomas 50
Gentner, Timothy Q. 3, 136, 136 n1, 158
Geoffroy Saint-Hilaire, Étienne xii, 29, 67, 81, 87, 89
Geoffroy Saint-Hilaire, Isidore 29–30
Ghiselin, Michael T. 82 n25, 86
Gibson, Robert M. 58, 73
Gilbert, Scott F. 109, 112–13, 113 n8, 114 n12
Godfrey-Smith, Peter 55
Goethe, Johann Wolfgang von xii, 64, 67
Golani, Ilan 76
Goldschmidt, Richard 29, 31
Goodman, Corey S. 115 n13
Goodman, Nelson 79 n23, 80 n24
Goodwin, Brian 110, 116
Gould, Stephen Jay 22, 59, 77, 87, 112, 115–16, 115 n14, 126–7, 132 n26
Gräff, Johannes 107 n15
Grahn, Jessica A. 101
Grassé, Pierre-Paul 169
Gray, Russell D. 114, 146 n9

Greene, Harry W. 65 n3, 69–71, 69 n7, 71 n10, 74, 76
Griesser, Michael 57, 143 n5
Griffiths, Paul E. 55–6, 110, 114
Grodzinsky, Yosef 102
Grohmann, Kleanthes K. 3
Grün, Rainer 156
Gunz, Philipp 157

Haesler, Sebastian 102 n8
Hakem, Razqallah 130
Hall, Brian K. 21–22, 52 n9, 65, 103, 109, 114, 114 n10, 115 n13
Hamburger, Viktor 111
Hansell, Mike 23, 67–8, 67 n5, 70, 72 n12, 73, 145–6, 145 n7, 146 n9, 151
Harcourt-Smith, W.E.H. 34
Harkins, Debra A. 149
Harpending, Henry C. 34
Harvey, Paul H. 131
Harzsch, Steffen 168
Hass, Joel 148, 151
Hauser, Marc D. 2–3, 21, 45, 51–4, 58, 92, 125, 135–8, 137 n2, 140, 143 n4, 158–9
Hawks, John 35
Heine, Bernd 15 n6
Hendrikse, Jesse L. 115
Henning, Willi 83
Henshilwood, Christopher S. 156
Herder, Johann Gottfried von 47
Herzfeld, Chris 147 n10, 149
Hetherington, Renée 34–5
Hinde, Robert A. 153
Ho, Yu-Chi 80 n24
Hobbes, Thomas 134
Hockett, Charles C. 21, 49
Hodos, William 69, 70 n8, 71
Hofman, Michel A. 131, 133
Homer, Steve 184
Hopcroft, John E. 95 n2, 184
Hornstein, Norbert 5
Horridge, G. Adrian 133
Hublin, Jean-Jacques 157
Hume, David 80 n24

Hung, Liang-Yu 130
Hunley, Keith 35 n6
Hunt, Gavin R. 146 n9
Hunter, John 29
Huxley, Julian 110, 132 n26

Iwaniuk, Andrew N. 146 n9

Jablonka, Eva 18, 114
Jackendoff, Ray 3, 19, 36, 111
Jenkins, Lyle 3
Jespersen, Otto 49
Johnston, Timothy D. 114
Jones, Clive G. 67 n5
Joshi, Aravind K. 95 n2, 98

Karten, Harvey J. 107
Kaskan, Peter M. 127–9, 129 n21, 131
Kauffman, Stuart 116
Keller, Evelyn Fox 114
Kelly, Scott A. 104–5
Kelso, J. A. Scott 22, 116–17
Kennedy, John S. 68 n6, 76, 78, 78 n21
Ketterson, Ellen D. 104
Khabbaz, Nabil 99 n5
King, Adam P. 6, 91, 91 n1
Kiparsky, Paul 15 n6
Kirby, Simon 19–20
Kitcher, Philip 132 n26
Klein, Richard G. 28, 156–157
Klingberg, Torkel 157 n24
Klopfer, Peter H. xii, 72, 72 n11, 72 n12, 78, 104
Kofron, Christopher P. 70
Kornack, David R. 129, 129 n21
Kosslyn, Stephen M. 151
Krause, Johannes 28, 155
Krebs, John R. 45 n2, 73–5, 78–9
Krink, Thiemo 169
Kuida, Keisuke 130
Kuno, Susumo 13
Kuteva, Tania 15 n6

La Mettrie, Julien Offray de 48
Labov, William 18–19, 19 n10, 20 n11

Lahr, M. M. 35
Lai, Cecilia S. 102 n8
Lamarck, Jean Baptiste de 88
Lamb, Marion J. 114
Langer, Jonas 131, 131 n22
Laubichler, Manfred D. 109
Lauder, George V. 70 n8, 83
Le Guyader, Hervé 106 n14, 166 n8
Lefebvre, Louis 146 n9
Leibniz, Gottfried Wilhelm 48
Lenneberg, Eric H. 3
Lestel, Dominique 147 n10, 149
Lewis, Harry R. 184
Lewis-Williams, David 156
Lewontin, Richard 26, 31, 33 n4, 40, 77, 132 n26
Lieberman, Philip 22, 100–2, 133
Liégeois, Frédérique 102 n8
Linde Medina, Marta 113, 113 n9
Longa, Víctor M. xii, 15, 113 n9, 114
Lorenz, Konrad 74–75, 80
Lorenzo, Guillermo xiii, 15, 37, 41, 49 n5, 53, 60–62, 60 n12, 81–82, 89, 104 n10, 147, 149 n13, 157 n24
Love, Alan C. 21, 61, 68, 89–90, 103
Lucretius 47 n3

Macready, William G. 80 n24
Madden, Joah 146 n9
Maienschein, Jane 109
Mansuy, Isabelle M. 107 n15
Marigo, Valeria 130
Marler, Peter 50, 50 n7, 143
Marr, David 150
Marx, Karl 47
Mattick, John S. 107 n15
Maynard Smith, John 74, 75 n16, 112 n5
Mayr, Ernst 24, 32, 54 n11, 82, 110
McBrearty, Sally 27, 27 n1, 153 n15
McKaye, Kenneth R. 58, 73
McKinney, Michael L. 127, 129, 131
McNamara, Kenneth 127
Mehler, Mark F. 107 n15
Mekel-Bobrov, Nitzan 130
Mellars, Paul 28, 156, 156 n23

Mercer, Alex 130
Michel, George F. 149
Miller, Geoffrey 49 n6
Miller, Julie E. 102 n8
Millikan, Ruth G. 77
Minelli, Alessandro xii, 18, 22, 39, 70, 70 n9, 82 n26, 85 n27, 87, 87 n29, 103, 105–6, 106 n14, 109, 110 n2, 114, 126 n20, 134, 137, 140, 161, 166–8, 166 n8
Mitani, John C. 50
Mithen, Steve 156–7
Møller, Anders P. 58
Möller, Ralf 171
Montaigne, Michel de 48
Moore, Allen J. 58, 73
Moore, David S. 9, 114
Morgan, Thomas H. 110–111
Moro, Andrea 102
Müller, Gerd B. 109, 113–14, 115 n13, 117 n17
Musso, Mariacristina 102

Nagel, Thomas 159–160
Negri, Toni xiv
Nevins, Andrew 9
Newman, Stuart A. 103
Nietzsche, Friedrich 47
Nisbet, Ian C. T. 58, 73
Niyogi, Partha 19
Nolan, Jr, Val 104
Northcutt, R. Glenn 87, 103
Nuño de la Rosa, Laura xii, 116 n16, 117

O'Donald, Peter 58, 73
O'Donnell, Timothy J. 158
Odell, Garret M. 119
Odling-Smee, F. John 40
Okanoya, Kazuo 93, 146 n9
Olson, Wendy M. 109, 114, 114 n10
Orr, H. Allen 32
Oster, George F. 110, 116, 118–19, 122
Ouattara, Karim 141
Owen, Richard xii, 29, 66–8, 70, 72 n12, 73 n14, 81–2, 82 n26, 85–7, 89, 106 n12, 149, 173
Oyama, Susan 22, 114

Packer, Craig 58
Papadimitriou, Christos H. 184
Parker, Geoffrey A. 74, 75 n16
Parker, Sue Taylor 115, 127, 129, 131
Paterson, Hugh E.H. 32
Penrose, Roger 162 n1, 165
Pepyne, David L. 80 n24
Perruchet, Pierre 143 n4, 158
Peschke, Klaus 73
Petersson, Karl Magnus 158
Petrie, Marion 58, 73
Pettigrew, John D. 147 n11
Pfeiffer, John E. 156
Piattelli-Palmarini, Massimo 25, 58–9, 102 n8, 110 n2, 114 n11, 132, 132 n26, 139
Pigliucci, Massimo 110 n1, 113, 113 n9, 132 n26
Pinel, John P. J. 156 n22
Pinker, Steven 3, 19, 111, 164 n4
Place, Ullin T. 78 n21
Pliny the Elder 48
Ploog, Dieter 144
Plotkin, Henry 111, 164 n4
Plutarch 48
Poeppel, David 4
Ponce de León, Marcia S. 157
Port, Robert F. 162 n1, 165
Premack, David 51 n8
Price, George R. 74
Proctor, Heather C. 70
Provine, William B. 110
Ptak, Susan 155
Pullum, Geoffrey K. 11, 158
Pulvermüller, Friedemann 102 n6
Pusey, Anne E. 58
Putnam, Hilary 91 n1, 165
Pylyshyn, Zenon W. 5–6, 8, 91 n1, 92, 162

Raff, Rudolf A. 109
Rakic, Pasko 129, 129 n21
Rampino, Michael R. 34
Rasskin-Gutman, Diego 113 n9, 116 n15, 125
Reid, Robert G. B. 22, 24, 33–5, 37–9, 62, 112, 126
Reid, Thomas 47

Reiner, Anton 132 n24
Reiss, John O. 116 n15, 117
Rendall, Drew 70
Rey, Arnaud 143 n4, 158
Rice, Sean H. 110 n1
Richards, Evelleen 29
Richards, Marc 12
Richards, Robert J. 87
Ridley, Matt 49 n6
Rieppel, Olivier 83
Ristad, Eric Sven 149, 165 n4
Robert, Jason S. 109, 114
Rochefort, Christele 102 n8
Roelink, Henk 130
Roessler, Erich 130
Rogers, James 158–9
Roth, V. Louise 21, 103, 106
Rousseau, Jean-Jacques 46–7, 47 n3
Rubel, Lee A. 166
Rupke, Nicolaas A. 82
Ruse, Michael 60
Russell, Edward S. 81
Ryle, Gilbert 76–7, 76 n19, 78 n21, 79

Salie, Rishard 130
Sanderson, M. J. 70
Sarpeshkar, Rahul 171, 171 n14
Saussure, Ferdinand de 9–10
Savage-Rumbaugh, E. Sue 51 n8
Schabes, Yves 98
Scheutz, Matthias 162 n1, 165
Schierwater, Bernd 106
Schmidt-Rhaesa, Andreas 125 n19, 166–7, 168 n11
Schoenauer, Norbert 152
Schram, Frederick R. 22, 105–6, 106 n14
Searle, John R. 60, 162, 165
Serres, Étienne 29
Seyfarth, Robert M. 55, 60, 141
Shannon, Claude 54 n10
Shimizu, Tom 107
Shtyrov, Yuri 102 n6
Shu, Yousheng 172
Shubin, Neil 87, 115 n14

Simon, Herbert 103
Simpson, George G. 66, 82, 110
Sinervo, Barry 104
Singer, Tatjana 107 n15
Skinner, Burrhus F. 72 n13, 76 n17, 77 n20, 80
Slabbekoorn, Hans Willem 143
Slack, Jonathan M. W. 21, 86 n28, 105–6, 106 n14
Smart, John J. C. 78 n21
Smith, Brian Cantwell 165
Smith, John W. 45, 55
Smith, Kathleen K. 117
Smith, Kenny 19–20
Smith, Linda B. 22, 116–17
Smolensky, Paul 165
Sober, Elliott 54 n11
Soffer, Olga 154 n18
Soressi, Marie 153 n15
Spanier, Bonnie B. 58
Spiteri, Elizabeth 155
Stevens, Peter F. 83
Stich, Stephen 2
Stotz, Karola 110
Strausfeld, H. J. 133
Striedter, Georg F. xii, 87, 103, 106, 106 n13, 125 n19, 128, 129 n21, 132–3, 132 n25, 135, 144, 167
Stringer, Chris 156
Struhsaker, Thomas T. 55, 141
Su, Bing 130
Surbey, Michele K. 111 n3

Takahashi, Kaori 102 n8
Tattersall, Glenn J. xiii, 38, 60
Teramitsu, Ikuko 60, 102 n8
Tesnière, Lucien 10
Theißen, Günter 31
Thelen, Esther 22, 116–17
Theraulaz, Guy 169–70
Thornhill, Randy 58, 73
Tinbergen, Niko 75–76, 80
Tooby, John 75, 80
Townsend, David J. 1, 18 n9,
Tracy, Joseph 149

Ujhelyi, Mária 49–51, 50 n7, 143 n4
Ullman, Jeffrey D. 95 n2, 184
Ullman, Michael T. 102 n7
Uriagereka, Juan xii, 14–15, 23, 28, 102 n8, 146–7, 154–5

Valentine, James W. 70
van Gelder, Timothy 162 n1, 165, 166 n7
van Heijningen, Caroline A. A. 158
Van Valen, Leigh M. 82 n25, 86
Vanhaeren, Marian 156 n20
Varzi, Achille C. 150
Vicq d'Azyr, Félix xii, 64, 67
Vijay-Shanker, K. 95 n2
Vollrath, Fritz 169

Waddington, Conrad H. 112 n5, 119, 120
Wagensberg, Jorge 110
Wagner, Andreas 115
Wagner, Günter P. 86, 89, 106, 134
Wake, David B. 21, 65, 86, 103
Wang, Yin-qiu 130
Weaver, Warren 54 n10
Webster, Gerry 116
Weir, David 95 n2, 98, 99 n5, 178
Wenzel, John W. 65 n3, 70 n9

West-Eberhard, Mary Jane 39–40, 86, 112
White, Stephanie A. 60, 102 n8
Wiener, Norbert 54 n10
Wilson, Edward O. 31, 74, 79, 169 n13
Wilson, Robert A. 16 n8, 86 n28
Wimberger, Peter H. 70
Wittgenstein, Ludwig 1, 7, 14, 14 n5, 24, 76, 76 n19
Wolpert, David H. 80 n24
Woods, C. Geoffrey 130
Wouters, Arno 61
Wright, Larry 60 n12
Wu, Chung-I 35
Wynn, Thomas 156 n23, 157 n24
Wynne-Edwards, Vero C. 58, 73

Young, Bruce A. 85, 106
Yu, Fuli 130

Zahavi, Amotz 45 n2, 58
Zahavi, Avishag 45 n2, 58
Zelditch, Miriam L. 126 n20
Zhang, Jianzhi 130
Zilhão, João 153–154, 153 n17, 156
Zuberbühler, Klaus 50 n7, 141 n3, 143
Zylberberg, Ariel 172 n15

General Index

3-D space 148–9, 149 n13

abnormality
   developmental 26, 29, 152–3
   organic 29, 31
   see also monstrosity
aboutness 91
acceleration 127–8 *see also* peramorphosis
accommodation
   behavioral 40
   phenotypic 39
acculturation 153
action
   explaining 67, 74 n15 *see also* behavior (causes of)
   pattern 67–9, 75–6, 151, 169–70 *see also* modal action pattern
active memory *see* working memory
activity 23, 39, 45, 61–2, 68, 70, 80, 88–91, 93, 100, 103–4, 108–9, 122, 131–3, 158, 173
adaptability 24, 33, 37–9, 62, 151
adaptation 18–19, 33, 36–8, 50, 60, 62, 68
adaptationism 20, 25, 30, 36–7, 54 n11, 57, 60 n12, 75, 77, 80, 90, 111, 132, 163 n2
ADCYAP1 130
adequacy 4, 71 n10, 72, 131, 161
   descriptive (of grammars) 96, 176–7, 184
aesthetic judgment 146 n9
Africa 28, 35, 153 n15, 156
   Sub-Saharan 28 n2, 34, 153 n16
agreement 5, 14–15, 14 n5, 15 n6, 17–18
analogy 72 n12, 81
anaphor 5, 13–14 , 17
anatomy 68, 100, 122, 131, 133
   comparative 65, 67
ancestor 156
   as an asymmetrical relation 83–6, 85 n27

common 49, 66, 69, 71, 82, 87 n29, 155, 157, 166
animal 66, 68, 72 n11, 73–5, 75 n16, 81, 87, 87 n29, 91, 106 n14, 112, 144, 146, 158–9, 166 n8, 167–8, 171, 173
   definition of 86 n28
   see also Metazoan
anthropomorphism 58, 78 n21, 108
apoptosis 130 *see also* cell death
ape *see* primates
appendages 81, 87, 89, 105
   paramorphs 87 n29
archeological record 24, 28
archetype
   ArcheType 0 95
   ArcheType 1 95
   ArcheType 2 95
   ArcheType 3 95
   computational 95, 95 n2, 95 n3, 99, 105, 110, 122
   as draft organ 105–6
   Owenian 81–2, 85, 106 n12, 122
area
   parieto-temporal 102, 102 n7
   X 102
art 17, 156
arthropods 86–7, 87 n29, 168
   coleoptera 168
   dictyoptera 168
   hymenoptera 168, 168 n12
ASPM 130
aural pattern recognition 158–9, 176 n2
autapomorphy 155
automaton 96 n4, 140, 162 n1, 174–5, 174 n1, 176, 183–4
   as an acceptor 174, 177–8
   deterministic pushdown 178 n4

embedded pushdown 99 n5, 178
finite-state 93, 95–7, 142, 159, 164, 169–71,
    174, 177–8, 177 n3, 178 n4
  as a generator 174, 176
  hierarchy 177–8
  linear-bounded 95, 99 n5, 177–8
  pushdown 17, 95, 98, 99 n5, 177–8,
    178 n4
  pushdown hierarchy 178
axis paramorphism 87 n29 *see also* appendage

basal component 22, 133 *see also* sequencer
basal ganglia 100–2, 102 n7, 106, 108, 127–8,
    131, 132 n24
  grammar 100, 102, 133
bat problem 159
beads 153–6, 154 n19
behavior 3–4, 27–8, 33, 35, 38–9, 62, 68 n6,
    69 n7, 71 n10, 73, 87–8, 90, 92, 101, 103–5,
    113, 117, 129, 140, 152, 157–9, 163–5,
    165 n5, 176
  adaptive 74
  altruistic 73
  causes of 70, 73, 77–80, 105, 105 n11, 162
  central cases 66 n4, 76
  comparative analysis of 48–9, 52, 55–7, 60,
    63, 67, 69–71, 72 n12, 107
  complex *see* cognitive higher capacities
  constricting 69, 69 n7, 71, 77
  construction 67–68, 70 n9, 145–6, 145 n7,
    145 n8, 151–2
  control mechanism of 78, 165, 170–1 *see
    also* disposition
  courtship 73
  intentional 91
  gaping 70
  goal oriented 91
  as a natural kind 44, 54 n11, 58, 65–7, 69,
    71–2, 72 n11, 74–6, 80–1, 88 *see also*
    behavioral category
  parental care 70 n9
  peripheral cases 66 n4, 76
  as a process 68, 78
  social 20 n11, 146 n9

stereotyped 65, 65 n3, 151
stimulus-driven 78–9, 151, 162–3
verbal 40–1, 44, 177
behavioral
  accomodation 40
  category 21–2, 67, 69–71, 72 n13
  ecology 50, 73–5 *see also* sociobiology
  isolation 32
  repertoire 39–40, 156
  vocabulary 69, 72–3, 76–7
behaviorism *see* psychology (behaviorist)
bifurcation 118–19, 133
big oh notation 181
  constant ($O(c)$) 181
  exponential ($O(c^n)$) 181
  factorial ($O(n!)$) 181
  linear ($O(n)$) 181
  logarithmic ($O(\log n)$) 181
  polynomial ($O(n^c)$) 181
bilaterian 167
biolinguistics 3–5, 7, 20
biology 1–2, 5, 7, 17, 31, 56, 60 n12, 66,
    70 n8, 76–7, 86, 86 n28, 111, 114, 116,
    137, 161, 166, 166 n8
  comparative 63, 65–6, 69, 87, 115, 122
  evolutionary developmental *see* evo devo
birds 38, 69–70, 71, 83, 87
  alarm signals 57, 143 n5
  bowers *see* bowerbirds
  construction activity/behavior 23, 145–6
  knotting 23
  nests 70, 145, 164
  songs 23, 47, 47 n3, 49, 56, 92–4, 96, 102,
    105–7, 122, 136, 142–3, 146 n9
  weaving *see* weaverbirds
bowerbirds 73
braid 148, 151, 158, 164
  word 148
  see also knot
brain 89, 91, 162, 167
  development 107 n15, 127–33, 129 n21,
    132 n25, 153, 144, 157
  evolution 5–6, 62, 122, 126, 131–2, 131 n23,
    132 n25, 161, 166–7

brain (cont.)
    organization 3, 6, 28, 31, 39, 92, 102, 102 n7, 106 n13, 127, 131-2, 132 n25, 135, 140, 157, 167-8, 179
    size 125 n19, 127, 129-31, 167
Broca's
    aphasia 101
    area 100-2
burials
    intentional 156
    hygienic 156, 156 n22
    symbolic 156

calls 49, 50 n7
    alarm 50 n7, 56-7, 141-2, 143 n5, 158
    contentful 143-4
    long 50
canalization 112 n5
Cartesianism 2, 4, 78 n21, 79
Case 5, 14-15, 14 n5, 15 n6, 17-18
CASP3 130
CDK5RAP2 130
cell 103, 107, 113-14, 114 n10, 116, 119, 126-7, 166, 180
    asymmetric division 128-9
    death 130
    glial 131
    migration 127
    neural 129
    precursor 127-8
    symmetric division 127, 129
CENPJ 130
central computational complex (CCC) 16-17, 20, 22-4, 27, 36, 42, 91-2, 108-9
    human (CCC_{HUMAN}) 22-5, 90, 102, 107, 109-11, 125, 132 n25, 133
Cercopithecidae 56-7, 141, 141 n3
cerebellum 101, 102 n7, 131
change 22, 29, 33, 35-9, 61-2, 99, 109, 112-17, 113 n6, 120, 125-6, 126 n20
    linguistic 15, 18-19, 19 n10
    of state 94
character 83-4, 87, 107, 111-12, 114, 126, 135, 147

behavioral 58, 74, 78
morphological 74
Châtelperronian 156 n23
Chomsky hierarchy 13, 22, 51, 94-9, 95 n2, 99 n5, 109-10, 122-125, 154, 162 n1, 173-4, 176-80
Chomsky normal form 175 nb
Chomskyan layer *see* long-distance relations (crossing)
Church-Turing thesis 162 n1
circuits
    cortico-subcortical-cortical 100-1, 108
    dorsolateral prefrontal 100-1
    intra-cephalic 107
cis-regulatory sequence 113, 113 n7 *see also* gene (regulatory)
cladistics 83
classes
    of complexity *see* complexity
    of equivalence 95 n3, 173, 182 n6
    natural *see* natural kind
classical architecture *see* computation (digital)
closed under complementation *see* set (recursive)
clothing 28, 154 n18
Cnidaria 161, 166-8, 167 n9, 168 n10
cognition 5, 22, 65-6, 88, 92, 100-1, 109-11, 122, 163 n2, 171
    animal 43, 48, 51-2, 133
    human 1-3, 28, 51, 92, 111, 133, 136, 157, 157 n24, 161
    nonhuman 3, 131, 136, 147, 159
    numerical 52, 92
    social 52, 92
cognitive
    higher capacities 65, 65 n3, 80, 153, 167-9
    module 5, 20, 92, 111, 163-5, 164 n4, 165 n5, 168-71
    science 66-7, 91 n1, 162 n1
communication
    ancestral system 48-9
    animal 4, 21, 23, 44, 46-8, 51-7, 51 n8, 60, 66

as behavior 48–50, 52, 55, 107
evolution of 45, 48, 49 n6, 52, 57
as an organic function 43, 44–5, 52–3, 54 n11, 56, 58
as an instinct 46, 47 n3
sematectonic *see* stigmergy
systems 21, 43, 44–5, 51–4
theory 54 n10
communicative
approach 15 n6, 18, 19 n10, 21–23, 25, 46–7, 51, 54, 54 n10
fallacy 44–6, 48–9, 51, 54, 57, 59
trascendentalism 57
complex dynamic systems theory 22, 116–17, 119, 165–6, 165 n7
complexity
computational 9, 23–4, 52, 96 n3, 108, 124, 146 n9, 158–159, 173, 176–7, 179–80, 183–4 *see also* problem
degrees 51–2, 95, 95 n2, 97, 99, 99 n5, 102, 136, 138
EXP (class) 181–3
hierarchy 148, 181, 183
lower bound *see also* problem (hard) 176, 181
measure 180–182
NP (class) 148–9, 151, 182–3
of algorithms 180
of cognitive abilities associated to technical abilities 23, 28, 145 n8 *see also* knot
of cognitive abilities associated to vocal abilities 50, 53, 96, 141, 143 n4
of knot tying 147 *see also* knot
of linguistic expressions 5–6, 8–9, 12, 28, 51, 97, 137, 142, 165 n4, 178
P (class) 181–3
PSPACE (class) 148–9, 182–3, 183 n7
space 182
structural 22, 94–5, 95 n2, 109–10, 122, 173, 174, 176, 180 *see also* Chomsky hierarchy
time 182
upper bound 181–182 *see also* problem (complete)

computation
analog 166–7, 171, 172 n14
animal 91
as an organic activity 90, 100, 122
deterministic mode 159, 169, 171, 178 n4, 180, 182
digital 91, 161, 165, 171, 171 n14
models 123, 173–4, 177 *see also* automaton; Turing machine
nondeterministic mode 93, 142, 178 n4, 180, 182
organ *see* computational system; *see* central computational complex
computational
activity 91, 100, 104, 109, 122, 133
approach 22–3, 91 n1 *see also* computationalism
mind 140
computational system (CS) 6–7, 16–17, 22–3, 51–2
diversification 22, 92, 95, 108, 124–5, 133, 138–40, 143–4, 154, 157, 164
human (CS_{HUMAN}) 137, 138–40, 157 n24
natural 91–2, 94, 100, 108, 126, 127, 139, 142, 144–5, 161, 164–5, 164 n4
of weaverbirds (CS_{WEAVER}) 145, 152
computationalism 162, 162 n1, 165
wide 16 n8
computers 75, 91, 169
compuype 21, 105–6, 122–4, 125 n19, 126, 129, 149
compuT0 123, 178–9
compuT1 105, 123, 138–9, 143, 149, 154–7
compuT2 105, 123, 143 n4, 154
compuT3 105, 123, 142–3, 178
conceptual-intentional systems *see* systems of thought
context-free 11, 95, 95 n2, 97, 109, 158, 175–8, 175 nb, 176 n2, 177 n3, 178 n4, 183–4
recognition *see* problem
context-sensitive 95, 95 n2, 109–10, 147–8, 164, 175, 175 na, 175 nb, 177–8, 183
mildly 95 n2, 99 n5, 110, 149, 176, 183 *see also* problem
recognition *see* problem

continuity 48–9, 51 n8, 83, 86, 87 n29, 138
  paradox 107, 125 see also language evolution
control theory see cybernetics
control unit 177–8
co-option 92
correlation patterns 89
cortex 102, 127–8, 132–3
  neo- 106, 129–30, 144
  prefrontal 100–101
cortical
  component 22, 100, 102, 122, 133 see also working memory
  invasion 157
  structure 129, 131, 133
cotton-top tamarins 158
counterfactuals 59, 132
courtship see seduction
creativity 151
ctenophores 166
cues 58, 168, 170
cybernetics 78, 78 n22
cytogenesis 127, 128 see also (precursor) cell

dance 58, 73
Darwinism 29–30, 45, 48–51, 49 n5, 54 n11, 80–1, 83, 88, 112 n5, 116, 124, 132
  neo- 57, 74–5, 110, 110 n2, 114
deception 56
decidability 99 n5
derivation 174–5, 177
descent 76, 84, 88, 139
development
  as creative force 110, 115–16
  as a dynamic non-linear process 110, 116
  as genetic program 114
  concerted 128, 131–2
  constraints on 89, 106, 110, 116, 132 n25, 138
  embryonic 89, 105, 106 n14, 127, 129, 133
  epigeneticist view 114, 119
  factors on 22, 42–3, 104, 106, 109–10, 113–16, 123–5, 134
  inertias in 87
  mosaic 131, 132 n25, 133

path 89, 112, 112 n5, 115, 120, 123, 125, 157
  perturbations of 22, 109–10, 115–17, 119–21, 124–6, 129, 133
  preformationist view 114
  primacy of 18
  systems of 30, 45, 104, 115–18, 165
developmental reprogramming/repatterning 115, 126 n20
developmental systems theory 22
display 45, 58, 69
disposition 67, 77–9, 79 n23
  genetic 79
  see also behavior (causes of)
domain specific 2–3, 5, 7, 20, 92
dorsal ventricular ridge 106, 106 n13
drive see disposition
dualism 78 n21, 104 n9
duets 50

echolocation 147
ecosystems ecology 67 n5
efficiency
  communicative 18, 45 n2
  computational 180–1
electrolocation
  in gymnotoid fishes 147
  in monotremes 147 n11
  in mormyrid fishes 147
embryology 111, 115 n14
emergence 22, 38–9, 112, 117
  critical-point 126
emotion 56, 100
  expression of 47–49
empiricism 88
environment 15, 16 n8, 33–4, 37–41, 61–2, 90, 111, 115, 163, 166, 169
  as a stimulus-generator 78 see also behavior (stimulus-driven)
epigenetic landscape 119
equivalence (of grammars)
  strong 176
  weak 176
  see also generative capacity

General Index 231

Eshkol-Wachman notation (EW) 76
essentialism 54 n11, 86, 86 n28
ethology 75, 161
evo devo 8, 22, 109–17, 110 n1, 110 n2, 113 n8, 115 n13, 115 n14, 121, 135
evolution 16 n8, 29–30, 37, 45, 59, 67 n5, 72, 75, 78, 83–4, 86–7, 90, 109–15, 114 n10, 124, 131–3, 135, 144–5
   concerted 132, 157
   morphological 22, 109, 116–32, 139
   mosaic 133
evolutionary normalization 26
evolutionary novelty *see* emergence
evolutionary trajectory 22, 29, 117–19, 133, 143
evolvability 115, 131 n23, 132
explanation
   adaptationist 20, 25, 30, 36–7, 50, 54 n11, 55, 57, 60 n12, 75, 77, 80, 90, 111, 132, 163 n2
   causal 37, 43, 59, 78
   externalist 111, 111 n4
   internalist 109, 111, 111 n4
   panglossian 77
   psychological 75–79
expressive power 53
external systems 7, 16–17, 23, 26–7, 91–2, 140, 143
external connections *see* external systems
externalization mechanisms 51–2

faculties 2, 137, 140, 164 n4, 171
   horizontal 164 n4
   vertical 163
faculty of language (FL) 1–20, 15 n6, 22–3, 26–8, 35–6, 39–41, 90, 92, 95 n2, 102, 107–8, 110–11, 134–40, 142–4, 153–7, 161 n1
   factorial/combinatorial character 2, 137, 140
   in a broad sense (FLB) 3, 51–2, 135, 138
   in a narrow sense (FLN) 3, 51–3, 92, 135–6, 138
   neanderthal ($FL_{Neander}$) 155
   uniqueness 20, 47, 51–2, 134–5, 138
fallacy
   communicative 44–6, 48–9, 51, 53–4, 59
   functionalist 44, 46, 57–8

family resemblance
   (peripheral cases) 76
   see also behavior (central cases); behavior
features 14, 16–17
   conditions of use 14
   non-interpretable 15, 16 n8
fight behavior 58
finches 60, 93, 96, 142
finite-state systems 52, 95–7, 109, 142, 158, 164, 170, 177–8
   deterministic 159, 169, 171
   non-deterministic 93, 142
fixed action pattern (FAP) 76 *see also* modal action pattern
foraging 146
form 30–1, 33, 57, 64, 67, 76, 82, 86, 94, 109, 113 n6, 159, 173
   primacy over function 37, 43, 60–62, 64–5, 76
formal language
   as a set of strings 158, 173–4, 176 n2
   theory 22, 96 n4, 122, 133, 136, 140, 176, 183–4
   see also regular system; context-free; context-sensitive; and unrestricted system
fossils 154
   DNA 155
FOXP2/FoxP2 45 n1, 102, 102 n8, 107, 155
function 14, 27, 33, 36, 40, 44–5, 57–8, 61, 63–73, 68 n6, 72 n12, 74 n15, 76, 80–1, 86, 94, 106 n14, 112, 126, 130, 159, 172–3
   adaptative 30
   as activity *see* activity
   behavior *see* use
   causal 55
   criteria for grounding 4, 70 n8, 80
   disembodied 37
   essential *see* function (special)
   organic 43, 45, 52
   parasitic 59–60, 126, 132
   proper 37, 54 n11, 77
   reproductive 49
   selected-effect concept 55, 60 n12
   special 38, 53, 59–60

function (*cont.*)
　structure *see* activity
　use 61, 68, 90, 105, 153
　words 50
functionability 62
functional
　architecture 5–6, 8, 92, 100
　specialization 53, 91
functionalism 30, 43, 172
　anti- 53
　transcendental 37, 57–8, 64

gametic incompatibility *see* reproductive barrier
geometric primitives 150
gene 22, 33, 102–4, 112–14, 113 n9, 114 n10, 114 n12, 119–120, 129–30, 166–7
　allelic variants 112
　regulatory 39, 107, 107 n15, 113, 113 n6, 129, 155
　target 113, 155
generation (of representations) 18 n9, 94
generative capacity
　strong 176–7, 176 n2, 184
　weak 176–7, 184
generative linguistics 96 n4, 176
genetics 155
　developmental 113, 113 n8
　population 24, 111, 113–14, 113 n7, 113 n8, 113 n8
　see also paleogenetics
genotype 15, 74, 75 n16
geometrical patterns 156 n21
gestures 46, 49, 150
　articulatory 144
gradualism, in evolutionary thought 30
grammar
　as cultural/conventional system 8–9, 12–19, 26–7, 35–6, 40–1, 42, 46
　as finite sets of rules 174
　as system of habits 16
　equivalence with automata 174, 176–8, 178 n4, 184
　hierarchy 51, 94–6, 95 n2, 99 n5, 109, 173–80

　as historical product 2, 8, 12, 14–19, 27, 40–1, 42
　as human invention 5, 7–8, 14, 24
　as socially/culturally transmitted 2, 12, 16, 24, 40
　type 0 95, 174–5 *see also* unrestricted system
　type 1 22–3, 95, 95 n2, 97, 99, 99 n5 *see also* context-sensitive
　type 2 51, 95, 97–8, 99 n5, 100, 170 *see also* context-free
　type 3 52, 95–7 *see also* regular system
grammaticalization 15 n6
granularity/incommensurability problem 4–7
great apes 147 n10, 149–50 *see also* primates
grooming 146
growth 29, 115–16, 133, 157
　dendritic 130–1
　exponential 128
　fetal 130
　linear 131
　rate of (of a complexity function) *see* big oh notation

handicap principle 45 n2
head 10–11, 14
　final 11, 17
　initial 11, 17
heterochrony 116–17, 126–7 *see also* developmental reprogramming/repatterning
heterotopy 126 n20 *see also* developmental reprogramming/repatterning
hierarchical organization
　of linguistic objects 10–12, 52
　in the structural analysis of nature 6, 21, 103, 105, 119, 138
hitches 146, 150–1
homeostatic equilibrium 53, 55–6, 79 *see also* neurophysiology
hominid 36, 49
homology
　ahistorical concept 81–2, 85, 87, 104
　behavioral 44, 65–7, 67 n5, 69, 71–72, 72 n11, 75–6, 81, 88

biological concept 89, 134
computational 21, 23, 89–105, 107, 138, 149, 173
   as correspondence 86
   Darwinian *see* homology (historical concept)
   deep 87, 87 n29
   developmental *see* homology (biological concept)
   general 66, 85, 149
   of activity *see* homology (computational)
   historical concept 66, 71, 72 n11, 74, 76, 82–4, 85 n27, 86–7, 87 n29
   Owenian *see* homology (ahistorical concept)
   serial 81 n25
   special 66, 85, 149
   structural 69, 71, 88, 137
   as a symmetrical relation 84
homoplasy 83, 87 *see also* analogy
human specificity 2–3, 11, 16, 26–7, 36, 46, 48, 136
humans
   anatomically modern 24, 28, 153–7, 153 n15, 153 n16, 154 n18
   early 27
   transitional archaic 28, 153 n15
Hume's Problem *see* learning (inductive)
hypercomputation 162 n1
hypermorphosis 127–8, 131 n22 *see also* peramorphosis

inflection 50
information 18–19, 19 n10, 54 n10, 59, 90, 157 n24, 166–8
inheritance 9, 66
   Mendelian-Morganian 112–114
   neo-Lamarckian 115
innateness 2
input
   string 177–8, 181
   tape 177–8
interface 10, 16–17, 52–3, 92, 108, 139, 143–4, 153, 163–4, 164 n4, 167

insects *see* arthropods
instinct 46, 79, 146, 151 *see also* learning
instruction *see* learning
invertebrates 73, 89
isolation
   behavioral 32
   ethological 32
   geographical 32, 33 n4
   sexual 32
   see also reproductive barrier
isomorphism 103–106

jump *see* emergence
just-so story 7, 24, 35

knot
   bowline 150
   butterfly 149
   mathematical theory 147–8
   overhand 146, 148
   real 151
   recognition *see also* problem (UNKNOTTING *and* GENUS) 150–1, 183
   sheepshank 149
   slip 146, 149
   topological properties 147–51, 149 n13, 158
   trefoil 148
   tying abilities in humans 146, 149, 154, 154 n18
   tying abilities in nonhuman animals 23, 145–7, 150, 164 *see also* birds
   unknot 147–51, 149 n13
know-how 67
   vs. know-that 67
key innovation *see* emergence

lamination 133
language
   as behavior 40
   change 18–19, 19 n10
   as a communication system 19 n10, 21, 43–54, 47 n3
   computational aspect of 151–2, 102 n7

language (cont.)
  concepts of 1–2, 40–1, 47, 158
  conventional 46
  dual nature of 5, 7
  evolution 4–5, 19, 21, 41, 42–3, 46,
    49 n6, 50, 136
  external 14 n5, 15, 21, 53
  formal *see* formal language
  games 14 n5
  minimal 50
  natural 46, 96–8, 105, 147, 149, 176, 183
  organic dimension 7, 24
  pathological view of change 19 n10
  polysynthetic 10
  as a problem *see* problem
  rules 8, 10–11, 13, 17, 159, 174–5, 175 na,
    175 nb, 177
  as a set of strings *see* formal language
  sign 10
  as a social institution 46
  social uses of 59
  specificity 3, 6, 8, 11, 16–17, 26,
    92, 136
  of thought hypothesis 162–3
late termination 22
learning 40, 79–80, 102, 146, 150, 169
  capabilities 79
  generalized 80 n24
  inductive 80 n24
  supervised 146
  theory of 67
  vocal 146 n9
level
  of biological analysis 103, 119, 160
  of computational analysis/complexity
    22–3, 50–2, 94–5, 95 n2, 97, 99, 99 n5, 102,
    105, 108–10, 122, 124, 136–138, 143 n4, 148,
    173–6, 179, 181–3
  hierarchical 6, 21, 103, 138, 173–4
  quasi-independence 103
  of structural analysis 21, 80, 103
limbs 68–9, 71, 81, 87, 87 n29, 89, 119 *see also*
  appendages
linear order 10–11, 96

linguistic
  conventions *see* grammar as cultural/
    conventional system
  mechanisms 4, 7, 26, 51–2
  representations 4, 7, 26, 51
  see also language (specificity)
linguistics 4, 7, 21, 49, 161
  evolutionary 3, 5, 7, 16, 136
logic of monsters 30
long-distance relations 12–14, 17, 99, 136
  crossing 12, 14, 99, 136

machine
  abstract *see* automaton
  digital *see* computation (digital)
  sequencing *see* computational system
  virtual *see* automaton
maladaptivity 19, 19 n10, 32–3, 111
mammals 69–71, 72 n11, 106–107, 127, 129, 172
marine isotope stage (MIS)
  MIS3 35
  MIS4 34–5
Markovian process 147
materialism *see* physicalism
MCPH1 130
memory 16 n8, 18, 97–8, 99 n5, 102,
  102 n7, 106–7, 110, 122–3, 156 n21,
  157 n24, 164 n4, 177–9
  regime 99, 133, 177
  see also working memory
mereological relations 150–1
mereotopological relations 151
Metazoan 70, 125 n19, 145, 163–4, 166
microtubule nucleation 130
mind 3, 5–6, 9, 11, 13, 16, 20, 23, 26–7, 60 n12,
  76 n19, 78 n21, 88, 92, 111, 133, 135–6,
  137 n2, 140, 159, 162–5
  architectural organization 8, 24, 111, 140,
    164–5, 165 n6
minimalism 5–6, 15, 15 n7
modal action pattern (MAP) 69–70,
  69 n7, 76
Modern Evolutionary Synthesis *see* Modern
  Synthesis

Modern Synthesis 110, 110 n2,
    111 n3, 112–14
modern cognitive toolkit 27
modularity 39
    massive 164 n4, 165 n6
    of mind 163, 164 n4
module
    Chomskyan 164 n4
    Fodorian 164–5, 165 n5, 170–1
    mental 5, 7, 92, 171
    specialized 20, 92, 111, 154, 163–5, 164 n4,
        168–9, 171
monkeys 129–31, 131 n22
    Campbell's 141–4, 143 n5, 158–9,
        164, 177
    Old World 50 n7, 55–7, 130
    rhesus 130
    vervet 60
monster 24–5, 28–9, 33 n4, 38
    biological 29–30
    cognitive 31–2, 36
    hopeful 31–3
    social 31, 31 n3
monstrosity 24, 31–2, 35
    as a mechanism of species
        transformation 29
    as the outcome of abnormal
        development 26–7
    predictability/regularity of 24 see also logic
        of monsters
morphological evolution 22, 109, 111, 116, 122,
    126, 139
morphospace
    computational 22, 123–5, 133
    developmental 22, 118, 121–5, 138, 143
    empirical 125
    theoretical 125
mushroom bodies 168–9, 168 n12
music 49, 94, 101
    notes 93–4
    traditions 156
motor control 5, 35, 100, 144, 150
    fine 146
    voluntary 144

mutation 19, 114 n10, 117
    point 102, 110, 112–13, 129, 133, 155
myelinization 130–131

natural kind 44, 46, 54, 54 n11, 56, 58, 64, 66,
    91, 93, 107, 122
navigation 52, 92, 168, 171
neanderthal 23, 135, 140, 153–4, 153 n15, 156–7,
    156 n23
    knotting abilities 28, 154, 158–9
    linguistic abilities 155–8
    necklace 154
neoteny 127 see also paedomorphosis
nervous system 103, 110, 127, 131, 145, 166–7,
    166 n8, 171
    central (CNS) 167–9, 173
    cephalized 167
    complex 90–1, 135, 140
    as information processing system 91, 166,
        172 n15
    net-like (plexus) 166
    metazoan 145, 166
    peripheral (PNS) 167
nest building
    in birds 70, 79–80, 145–6, 145 n7, 145 n8,
        151–3, 164
    in social insects 169
    in termites 169–70
    in wasps 70 n9, 75, 170–1
    see also birds (construction/activity
        behavior)
nested embedding see hierarchical
    organization
    unlimited 53, 136
neural tube 127, 130
neurobiology 76
    developmental 127
neurogenesis see cell
neurophysiology 80
    as a response-generator 78
neuroscience 4
niche 33, 40–1
nomic sensitive creatures (NSC)
    164, 164 n3, 171 see also property (nomic)

nonnomic sensitive ceatures (NNSC) 163–164, 168 n10, 171 *see also* property (nonnomic)
noise 47, 48 n4

object recognition 150–1
  as parsing 150
ochre 156 n21 *see also* pigment
ontogenetic path *see* development (path)
ontogeny 78, 80, 109, 115 n14, 161 *see also* development
operative memory *see* working memory
organ 2, 22, 24, 47, 68, 81, 82 n26, 86, 89, 94, 105, 109, 119, 121–5, 131, 133, 139, 146, 159, 161, 167–8
  factorial/combinatorial character 39, 134, 137, 140
other people problem 159
overdevelopment 13, 28, 127, 131, 144 *see also* peramorphosis

paedomorphosis 127 *see also* heterochrony
paleoanthropological evidence 5, 28, 35, 153–4, 153 n15, 156
paleogenetics 28, 155
pallium 106, 106 n13, 108
paramecia 161–3, 165 n5
parameter
  control 117, 119, 129
  developmental 111, 114, 116, 119, 122, 133
  morphogenetic 109–10, 116–17, 123–5, 133, 138
parametric space
  cognitive 110
  as developmental constraint 110, 110 n1, 117, 119, 121–2
  see also morphospace
parcellation 133, 157
part-whole relations *see* mereological relations
partial reference terms 56
Passeriformes 145
  Emberizinae *see* weaverbirds (New World)
  Fringillidae 145
Icterini 145
Ploceidae *see* weaverbirds (Old World)
pattern generator *see* sequencer
peramorphosis 127–8 *see also* heterochrony
peripheral systems *see* external systems
phase shift 119, 133
phenogrammatical 12 n4
phenotype 4, 9, 22, 34, 39, 54 n11, 60, 74 n15, 90, 95, 113, 113 n9, 114 n10, 117–21, 123–4
  actual 125
  cognitive 18, 21, 33, 111
  computational *see* computype
  extended 18
  impossible 125
  landscape 110 n1
  possible 125
  potential 125
pheromones 58, 73, 170
phonemic inventories 35
phylogeny 67 n5, 70–1, 70 n9, 75, 80, 109, 115 n14, 116, 161, 163
phylotype 21, 105, 106, 106 n14
  arthrotype 106
  cyclotype 106
  malacotype 106
  phyloT1 106
  phyloT2 106
  phyloT3 106
  platytype 106
  trimerotype 106
phylotypic stage 105
physicalism 78 n21
pigment 27, 153–4, 153 n15
pigmentation 58
plasticity 39–40, 62, 80, 169
  developmental 39
Platonism 37, 82
Pleistocene 34
polynomial witness *see* succinct certificate
population
  bottleneck 33–5, 35 n5
  founding 24
  thinking 54 n11

postdisplacement 127 *see also* paedomorphosis
predisplacement 127, 131 n22 *see also* peramorphosis
predisposition *see* disposition (genetic)
primates
　higher 130
　nonhuman 23, 49–51, 50 n7, 51 n8, 127–31, 137 n2, 143 n4
principle
　of compositionality 141
　of maximal congruence 106
problem
　complete 149, 182–3, 182 n6, 183 n7
　CONTEXT-FREE RECOGNITION 183
　CONTEXT-SENSITIVE RECOGNITION 183
　decidable 179–80
　decision 175–6, 179
　as a formal language 180
　GENUS 148
　hard 149, 180–2
　MILD-CONTEXT-SENSITIVE RECOGNITION 182
　solving 101, 181
　UNKNOTTING 148–9, 151
　unsolvable 180
procedural memory *see* working memory
processing 11–13, 16, 18, 18 n9, 28, 98, 102 n6, 149, 164–72, 172 n14
progenesis 127 *see also* paedomorphosis
pronoun 5
property
　nomic 163–4, 167, 170–1
　nonnomic 163, 170–171 *see also* representation (mental)
psychology 56
　behaviorist 66, 66 n4, 72
　evolutionary 49 n6, 75, 90, 111, 114 n11, 164 n4

Quaternary 34

recursion 3, 52–3, 92, 136–7, 136 n1, 137 n2, 139
　tail 93, 96

trivial *see* recursion (tail)
reduction 69 n7, 74–5, 74 n15, 77, 78 n21, 113 n9, 182 n6
regular system 95–6, 142, 158, 175, 177, 178 n4, 184
　left-linear 175
　right-linear 175
　see also finite-state system
representation
　of knots 150–2, 164 *see also* mereotopological relations
　mental 4, 7–8, 26, 51, 56, 58, 91–2, 94, 143, 161–5, 163 n2, 167 n9, 168 n10, 169–71
　symbolic 155
representational theory of mind 91 n1, 162
reproductive barrier 32–3
resources
　cognitive 6–7, 9–10, 13, 19, 105, 107, 110, 147, 164 n4
　developmental 44, 87, 87 n29, 89, 138
　time and space 123, 180–2
　see also memory
response 33, 37, 39, 71 n10, 72 n11, 78, 80, 112, 162, 170
　displace or delayed 78 *see also* neurophysiology
rhythm 101
rules
　constraints on 13, 17, 93–4, 96, 174–5, 177
　format of 174, 175 nb

saltationism 30, 36
Saussurean layer *see* linear order
scaffolding 15, 19 n10, 166 n8
Scheherazade effect 49 n6
schismogenesis, complementary 33
second factor *see* environment
seduction 49, 56–8
selection
　as creative force 110, 112
　as filter 110, 116, 122
　natural 19–20, 30, 45, 48, 55, 60 n12, 64, 70 n8, 77, 80, 105, 110, 112, 112 n5, 114 n10, 115–16, 121, 124, 131–2

General Index　237

selection (cont.)
  sexual 49 n6
  strong positive (SPS) 129–30
semantic(s) 5, 54 n10, 73 n14, 92, 141, 143, 163
  thesis 52
sensitive period 14
sentence 14, 98–101, 105, 143
sequencer 22, 100, 102 n7, 133
set
  recursive 95 n2, 175, 175 na, 179
  recursively enumerable 175, 179–180
  see also formal language
shells
  intentionally perforated 28 n2, 153 n16, 155, 156 n20
  perforated by natural causes 153–156, 156 n20
  pigment-stained 153–4
SHH 130
Siberian jay 57
signal 21, 23–4, 45, 45 n2, 54 n10, 55, 57, 60, 166
  combined 142
  isolated 141
sociobiology 31, 50, 73, 75
spatial representation 56, 150–2
speciation
  allopatric 33 n4
  sympatric 24, 32
speech 10, 15–16, 18, 40–1, 100–1, 126, 134, 144
spinal cord 101, 127–8, 144
stack
  additional 99, 99 n5
  pushdown 98, 99 n5, 123, 178
starlings 136 n1, 158
stigmergy 165 n5, 169–70, 169 n13
stimulus 53, 78–9, 151, 162, 170
  external 78, 171 see also environment
  internal 79
  response model 78 see also psychology (behaviorist)
storage 99 n5, 178–9

strategy 74–5, 75 n16
  evolutionary 37
  evolutionary stable (ESS) 74–75
strings 13, 96–8, 155, 158–9, 175–81, 176 n2
  left-hand 175
  right-hand 175
  in mathematics 173–4
  sub- 97, 174, 175 na, 175 nb
structural description 97, 176–7
structuralism 43
structure
  dynamic dimension see activity
  linguistic 10–13, 12 n4, 18 n9, 51, 97, 98, 176
  morphological 103–106
  organic 30, 37–41, 45, 57–8, 60–63, 64–81, 89, 91, 103–4
  phrase 8
  word 8
subcortical
  structure 129, 131, 144
  see also sequencer
succinct certificate 182 see also complexity class, (NP)
supervenience 103
symbol
  manipulation 53, 91, 94–5
  nonterminals 174–5, 175 na, 175 nb, 177
  terminals 96–8, 174–5, 175 nb, 177
symbolic culture 28, 35
synapomorphy 155
synapse
  bidirectional 167
  chemical 167
  electrical 166–7
  unidirectional 167
synaptogenesis 131
syntax 20 n11, 50, 51 n8
  basic inferential relations 92
  lexical 50 n7
  minimal 50, 143 n4
  phonological 50 n7
syntactic engine see computational system
systems of thought 15, 136, 143–4, 154

General Index   239

technology 17, 27 n1
tectogrammatical 12 n4
teratology see monstrosity
territoriality 58, 73
Tesnièrean layer see hierarchical
    organization of linguistic objects
Thermoregulation 60
threading 154, 156
Toba (Sumatra) 34
tools see technology
topology 83, 147–8, 151
    equivalence 147–8
toucan 38, 60
Tower of London test 101
trace 14
transcription factor 102, 107
transitions between automaton states 93–4, 96
Turing machine 95, 105, 178–80, 182

underdevelopment 127 see also
    paedomorphosis
unity of type 82 see also archetype and
    ancestor (common)
unrestricted system 95, 175
Upper Paleolithic 156
use
    conditions of 14–15, 14 n5
    of forms 14 n5
    of language 14 n5, 18, 58
variation
    computational 108–9, 122–3, 138, 163–4
        see also computational system
    interspecific 5, 64 n1, 87 n29, 90, 106, 108,
        113–14, 135, 141

linguistic 8, 12, 18–19
in natural communication systems 52
phenotypic 114, 117–18, 122–5, 127,
    131, 163
vertebrates 3, 6, 22–3, 73, 73 n14, 76, 81, 82
    n26, 86, 87 n29, 89, 92, 126, 135, 167
vision 5, 150
    correlation with cephalized CNS 168
    in arthropods 168 see also mushroom
        bodies
    in Metazoan phyla 168
    see also object recognition
vocal-auditory channel see externalization
    mechanisms
voice 47, 47 n3, 48–9, 48 n4

weaverbirds 146–7, 151–2, 158, 164, 177
    New World 145
    Old World 145
well of attraction 119 see also epigenetic
    landscape
working memory 11, 13, 17, 22, 100, 102,
    102 n7, 105–6, 178
    accessible to conscious awareness 157 n24
    as attention 157 n24
    enhanced 157 n24
    networks 102
    subattentional 157 n24
    at a subconscious level 157 n24
workspace see working memory

zootype 21, 106–7, 106 n14
zooT1 107
zooT2 107
zooT3 107

The manufacturer's authorised representative in the EU for product safety is Oxford University Press España S.A. of el Parque Empresarial San Fernando de Henares, Avenida de Castilla, 2 – 28830 Madrid (www.oup.es/en or product. safety@oup.com). OUP España S.A. also acts as importer into Spain of products made by the manufacturer.

www.ingramcontent.com/pod-product-compliance
Lightning Source LLC
LaVergne TN
LVHW010339260326
834688LV00036B/786